Studies in Fuzziness and Soft Computing

Volume 303

Series Editor

Janusz Kacprzyk, Warsaw, Poland

For further volumes:
http://www.springer.com/series/2941

The series "Studies in Fuzziness and Soft Computing" contains publications on various topics in the area of soft computing, which include fuzzy sets, rough sets, neural networks, evolutionary computation, probabilistic and evidential reasoning, multi-valued logic, and related fields. The publications within "Studies in Fuzziness and Soft Computing" are primarily monographs and edited volumes. They cover significant recent developments in the field, both of a foundational and applicable character. An important feature of the series is its short publication time and world-wide distribution. This permits a rapid and broad dissemination of research results. Contact the series editor by e-mail: kacprzyk@ibspan.waw.pl.

Ajeet Kumar Pandey · Neeraj Kumar Goyal

Early Software Reliability Prediction

A Fuzzy Logic Approach

 Springer

Ajeet Kumar Pandey
AECOM
Hyderabad, Andhra Pradesh
India

Neeraj Kumar Goyal
Reliability Engineering Centre
Indian Institute of Technology Kharagpur
Kharagpur, West Bengal
India

ISSN 1434-9922   ISSN 1860-0808 (electronic)
ISBN 978-81-322-1742-8   ISBN 978-81-322-1176-1 (eBook)
DOI 10.1007/978-81-322-1176-1
Springer New Delhi Heidelberg New York Dordrecht London

Printed on acid-free paper

Springer is part of Springer Science+Business Media (www.springer.com)

Late Prof. R. B. Misra

Preface

Due to continuous growth in size and complexity of software system, reliability assurance becomes tough for developers while satisfying user expectations. Applicability of software keeps on increasing for products/systems ranging from basic home appliances to safety critical business applications. Development of reliable software with acceptable level of quality within available time frame and budget has become a challenging objective. This objective could be achieved to some extent through early prediction of number of faults present in the software. Early prediction of faults provides an opportunity to make early corrections during development process and reduces the cost of development. Further it also helps in better planning and utilization of development resources. This book is an upshot of our research at Reliability Engineering Center IIT Kharagpur, India. During our research we struggled a lot to find the way to predict software reliability at early stages of software development.

The book is intended to serve three associated audiences: software professionals, project managers, and researchers. For software professionals, this book serves as a means to understand reliability of their product, the potential factors affecting it, and adoption of best practices to reduce number of errors in the software. For software project manager, this book serves as a collection of tools, which can be easily modeled and used for monitoring reliability of software throughout the development process. Further, the reliability information obtained can be used for reducing software development time and cost. For software reliability and quality researcher, this book provides important references in the area. Besides, the models proposed in this book are basic framework. This framework can be put to critical investigation to find better methods which remain easy to understand, implement, and use. We hope that this book will help software professionals and researchers to solve complicated problems in the area of early software reliability with ease.

The book is divided into seven chapters. Chapter 1 starts with a discussion on the need of reliability and quality software. By providing various standard definitions of software reliability, error, fault, and failure, chapter has discussed two approaches for measuring reliability of a software system. First, a developer's view focusing on software faults; if the developer has grounds for believing that the software is relatively fault-free, then the system is assumed to be reliable.

Second, a user-based view which defines software reliability as "the probability failure-free software operation for a specified period of time in a specified environment". Irrespective of these viewpoints, software reliability measurement includes two types of activities: estimation and prediction. This chapter also discusses about the software reliability growth model (SRGMs) and its limitations. Chapter has highlighted the need of early software reliability prediction, its practical challenges, and solutions. The reliability relevant metrics and Capability Maturity Model (CMM) are also discussed in the chapter.

Chapter 2 provides background on software reliability and quality predictions. This chapter has focused on the literature surveys on software reliability models, reliability relevant software metrics, software capability maturity models, software defect prediction model, software quality prediction models, regression testing, software, and operational profile.

Chapter 3 proposes an early fault prediction model using software metrics and process maturity. For this, various reliability relevant software metrics applicable to different phase of software development are collected along with process maturity. Model approach is modular in the sense that it can be applied at each phase of Software Development Life Cycle (SDLC) and suitable for all type of software projects.

Chapter 4 has focused on fault density which seems to be a more desirable measure than number of fault for reliability and quality indicator. Fault density metric can be found for requirement document, design document, code document, and also for test report. The chapter proposes a multistage fault prediction model which predicts the number of residual faults present in the software as software industries are more interested to know residual faults present in their software.

Chapter 5 presents an approach to classify software modules as fault-prone (FP) and not fault-prone (NFP). Besides, the approach also predicts degree of fault proneness on a scale of zero to one, and ranks the module accordingly. ID3 algorithm along with fuzzy inference system is used for classifying software modules as FP or NFP. ID3 algorithm provides the required rules by analyzing history data for proposed fuzzy inference system.

Chapter 6 discusses an integrated and cost-effective approach to test prioritization that increases the test-suite's fault detection rate. The proposed approach considers the three important factors, program change level (PCL), test-suite change level (TCL) and test-suite size (TS), before applying any techniques to execute test-cases. These factors can be derived by using the information from the modified program version. A cost-effective reliability centric test case prioritization is also discussed in this chapter.

Chapter 7 discusses the way to develop the operational profile of a software system, to compute the number of test case required, and the way to allocate them. The chapter explains how the reliability of a software-based product depends on its operational profile. If developed early, an operational profile may be used to prioritize the development process, so that more resources are put on the most important operations. A case study using the traditional and operational profile-based testing is also discussed.

Appendices containing the data set used in various chapters of this are given the end. Readers are advised to synchronize with these dataset while utilizing the model for reliability prediction of any software or module. Any forms of constructive suggestions to improve our existing models are welcome. Readers can mail their queries/suggestions at ajeet.mnnit@gmail.com.

A. K. Pandey
N. K. Goyal

Acknowledgments

Writing the acknowledgment is the loveliest and toughest part. This book has been an outcome of our research at IIT Kharagpur and Cognizant Technology Solutions (CTS), Hyderabad, India. I am indebted to many persons from these organizations who directly or indirectly helped us during the preparation of the manuscript. It is my pleasure to acknowledge their help.

First, I would like to thank my professors, Prof. V. N. Naikan, Prof. S. K. Chaturvedi, and Prof. N. K. Goyal who not only introduced the subject to me but helped me a lot to explore the new opportunities and challenges of reliability engineering. I express my special thanks to Late Prof. R. B. Misra; he motivated me to start the journey of software reliability.

I acknowledge the help and cooperation received from Prof. P. K. J. Mahapatra, Prof. V. N. Giri, and Prof. U. C. Gupta during my stay at IIT Kharagpur. Thanks are also due to our EMS CoE coworkers of CTS, Hyderabad. I express my special thanks to Dr. Phanibhushan Sistu, Vivek Diwanji, and Smith Jessy, for their constant motivation and support. Other friends, who reviewed the manuscript and made helpful suggestions, are Abhisek Gupta, Ravi Kumar Mutya, Srinivas Panchangam, and Nitin Jain of CTS Hyderabad, Rakesh Chandra Verma of TCS Chennai, Prof. Rajesh Mishra of Gautam Buddha University, Greater Noida, Prof. Vivek Srivastava of NIT Delhi. Thanks are due to Prof. D. K. Yadav of NIT Jamshedpur, Prof. P. K. Kapur of Amity University, Noida, and Prof. M. M. Gore of MNNIT, Allahabad. I wish to thank Praveen Goyal, Director (Technical) at AECOM, Hyderabad; Dr. V. S. K. Reddy and Dr. D. Ramesh of MRCET, Hyderabad for their decent support during the last hours to finalize the manuscript on time. I also wish to thank Aninda Bose of Springer for his enthusiasm, patient, and support since beginning of the book, as well as Suguna Ramalingam, Production Editor (Springer), and his staff for their meticulous effort regarding production. I gratefully acknowledge the anonymous reviewers for their constructive comments and suggestions to improve this book significantly.

Special acknowledgment to MHRD, Government of India and administration of CTS, Hyderabad, India, for providing financial and administrative support during the draft of this book.

Finally, I wish to express my sincere gratitude to my parents Dr. D. P. Pandey and Kalawati Pandey, my parents-in-law Dr. M. L. Trivedi and Santosh Trivedi for their *AASHIRVAD*. I would like to acknowledge the continuous support provided by my wife Shweta and two kids Avichal and Khyati, whose laughter used to absorb my tiredness and let me complete this manuscript successfully.

Finally, I would like to wind up by paying my heartfelt thanks and prayers to the Almighty, for his love and grace.

A. K. Pandey

Contents

Acronyms

LOC	McCabe's line count of code
CC	McCabe's cyclomatic complexity
EC	McCabe's essential complexity
DC	McCabe's design complexity
EL	Halstead's executable line count
CL	Halstead's comments line
BL	Halstead's blank lines
CCL	Code and comment line
n1	unique operators
n2	unique operand
N1	Total operators
N2	Total operands
BC	Branch counts
FP	Fault-prone (goal metric)

Notation and Abbreviations

APFD	Average percentage of faults detected
CM	Coding metrics
CMM	Capability maturity model
DM	Design metrics
FCP	Number of faults at the end of coding phase
FDC	Fault density indicator at the end of coding phase
FDD	Fault density indicator at the end of design phase
FDP	Number of faults at the end of design phase
FDR	Fault density indicator at the end of requirement phase
FDT	Fault density indicator at the end of testing phase
FIS	Fuzzy inference system
FC	Functional complexity
FP	Function point
FP Module	Fault-prone module
FRP	Number of faults at the end of requirement phase
ID3	Interactive Dichotomizer 3
NFP Module	Not fault-prone
OP	Operational profile
OPBT	Operational profile-based testing
OEM	Original equipment manufacturer
PCL	Program change level
RC	Requirement complexity
RTS	Regression test selection
RTM	Regression test metrics
RM	Requirements metrics
RRSMs	Reliability relevant software metrics
SRGMs	Software reliability growth models
TCP	Test case prioritization
TSM	Test-suite minimization
TCL	Test-suite change level
TS	Test-suite size
TFN	Triangular fuzzy numbers
TM	Testing metrics

Chapter 1
Introduction

1.1 Need for Reliable and Quality Software

Nowadays, software is playing an ever increasing role in our daily lives, from listening music at homes to uninterrupted entertainment during travel, from driving car to ensuring safe air travel, and from variety of home appliances to safety critical medical equipments. It is virtually impossible to conduct many day-to-day activities without the aid of computer systems controlled by software. As more reliance is placed on these software systems, it is essential that they operate reliably. Failure to do so can result in high monetary, property, or human losses. Early sofware reliability prediction can help the developers to produce reliable software in lesser cost and time.

Software systems have become so essential to human in their daily lives that today it is difficult to imagine living without devices controlled by software. Due to this inevitable dependency on software systems, more attention is required during their development so that they operate in a reliable manner. Development of reliable software is challenging as system engineers have to deal with a large number of conflicting requirements such as cost, time, reliability, safety, maintainability, and many more. It has been found that the most of the software development activity is performed in labor-intensive way. This may introduce various faults across the development, causing failures in near future. The impact of these failures ranges from marginal to catastrophic consequences. Also, in most of the cases, both the cost of software development and losses from its failures are expensive. Therefore, there is a growing need to ensure reliability of these software systems as early as possible. Moreover, it is well known that earlier a fault is identified, the better and more cost-effectively it can be fixed. Therefore, there is a need to predict these software faults across the stages of software development process to achieve reliable software with cost-effectiveness.

A. K. Pandey and N. K. Goyal, *Early Software Reliability Prediction*,
Studies in Fuzziness and Soft Computing 303, DOI: 10.1007/978-81-322-1176-1_1,
© Springer India 2013

1.2 Software Reliability

Software reliability is defined as the probability of failure-free software operation for a specified period of time in a specified environment (IEEE 1991). In other words, software reliability corresponds to the successful execution of software operation for a specified period of time under specified conditions.

Software quality is the degree to which software possesses desired attributes in meeting laid down specifications. In this scene, software quality encompasses the array of software attributes such as usability, testability, portability, maintainability, and scalability. Software reliability is an important facet of software quality which quantifies the operational profile and life of a software system.

1.2.1 Software Error, Fault, and Failure

The terms error, fault, and failure are often used interchangeably, but these do have different meanings. According to the IEEE (1991) standard, an error is a human action that produces an incorrect result. For example, an incorrect action on the part of a programmer or operator may result coding error, communication error, etc. Fault is an incorrect step, process, or data definition in a computer program. A fault is the materialization of an error in the code. It is a software defect that potentially causes failure when executed. Alternatively, a fault is the defect in the program that, when executed under particular conditions, causes a failures. Also, fault is a developer-oriented concept and considered as a property of the program rather than a property of its execution or behavior. Faults may arise across the various stages of software development yielding requirements faults, design faults, and coding faults.

A failure is a departure of system behavior from user expectation during execution. A software failure is said to occur when the user perceives that it deviates from its expected behavior. Then, the software reliability is determined as the probability of software failure for a specified period of time in a specified environment. Hence, the software failure and reliability are more close to end user than developer as shown in Fig. 1.1. Potential failures found by the users or developers are traced back to find the root causes such as requirement faults or design faults or coding faults. It is the developer who identifies the root cause of these failures by tracing back to program design or code and find associated faults and errors. The reliability is usually influenced by finding and fixing these faults that causes failures. Thus, software error, fault, failure, and reliability have different views of software developers, testers, and users point of view. Figure 1.1 shows the relationship among error, fault, failure, and reliability as well as their association with software developer and end user.

As shown in the Fig. 1.1, an error results fault which is responsible for causing failure upon execution. The reliability of a system can be determined by finding

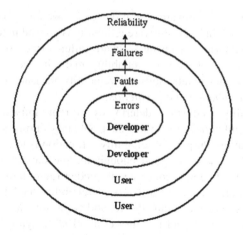

Fig. 1.1 Error, fault, and failure

the probability of these failures. Also, error and faults are more closely related to developer and therefore are placed in the circle as shown in the Fig. 1.1. Similarly, software failure and reliability are more closely related to end users and hence are kept in the same circle in the Fig. 1.1. In brief, error causes faults and a fault causes failures which are responsible of affecting software reliability. In other words, a fault is created when human makes an error. Errors may be in the requirement document, in the design document, in the code, or in the testing process causing faults in the developed software. These faults cause failures when got executed. Failures may be categorized as transient, permanent, recoverable, unrecoverable, non-corrupting, and corrupting based on their effect. Modeling and analysis of these software failures are important in software reliability measurement.

1.2.2 Measuring Software Reliability

Software reliability can be viewed differently by different people in different situations. A software developer, who views code as an instruction to hardware, may consider them reliable if each software requirement is executed properly by a related set of those instructions. A user sees the same software as a set of function and considers it reliable if it fulfills functional requirements without failing. Although, viewpoints are different, but there is a common agreement that software faults are the identifiable root of the reliability problems.

There are two approaches for measuring reliability of a software system. First, a developer-based view, focuses on software faults; if the developer has grounds for believing that the software is relatively fault-free, then the system is assumed to be reliable. Secondly, a user-based view which defines software reliability as "the probability of failure-free software operation for a specified period of time in a specified environment." In brief, software reliability is viewed in terms of software faults (the developer's view), causing software failures (the user's view). These two different viewpoints of software reliability make its measurement difficult. Irrespective of these viewpoints, software reliability measurement includes two types of activities: *estimation* and *prediction*.

Software reliability estimation determines current software reliability by applying statistical inferences to failure data obtained during system testing or during system operation. Purpose of reliability estimation is to determine whether the applied reliability model is good enough for the current software project. Software reliability prediction is concerned to predict future failure behavior from present and past failure data. Several software reliability models have been proposed for estimating and predicting the reliability of software systems. Details about these models can be found in Musa et al. (1987), Lyu (1996), and Pham (2006). Software reliability models provide quantitative measures of the reliability of software using failure data collected during testing or operation.

A software reliability growth model (SRGM) assumes that reliability of the software will continue to grow if observed errors during testing are removed. In other words, the number of residual software errors decreases with progress of software testing. SRGMs are useful only if failure data are available. Alternatively, SRGMs describe the behavior of software failures with respect to time and are applicable during the testing phase of software development. Thus, depending on the availability of failure data and software development stage, reliability prediction involves different techniques:

1. When failure data are available (e.g., software is at test or operation stage), reliability can be predicted using various SRGMs. Details about various SRGMs can found in Jelenski and Moranda (1972), Shooman (1972), Littlewood and Verrall (1973), Musa (1975), Schick and Wolvertion (1978), Goel and Okumoto (1979), Yamada et al. (1983), Kapur and Garg (1990), and Chatterjee et al. (1997).
2. When failure data are not available (e.g., software is at the design or coding stage), SRGMs cannot be applied and therefore researchers had to use fault statistics obtained during technical review of requirement, design, and code to predict software reliability (Gaffney and Pietrolewiez 1990). These metrics can be derived from different phases during software development. Predicting reliability of software system before testing in the absence of failure data is termed as early software reliability prediction (Gaffney and Pietrolewiez 1990; Agresti and Evanco 1992; Rome Laboratory 1992).

1.2.3 Limitation of Software Reliability Models

Around hundreds of SRGMs have been developed to date. One of the practical problems with these models is that these models require failure data (either test failure or field failure) for reliability prediction. Another problem with SRGMs is that they follow black-box approach for reliability prediction by considering system as a whole. SRGMs do not consider the factors such as internal structure of software, development process characteristics, product characteristics, and nature of human errors occurring during software development.

Predicting reliability of software before testing in the absence of failure of failure data are known as early software reliability prediction. Early reliability prediction attracts both software developers as well as managers as it provides an opportunity for the early identification of software quality, cost overrun, and optimal development strategies. Problem with early software reliability prediction is twofold:

- First, how to find software failure intensity function without executing the software, which is required to calculate the software reliability?
- Second, how time parameter of reliability evaluation can be found during early stage of software development?

Due to these constraints, failure behavior of software on time basis is very difficult to predict.

1.3 Why Fault Prediction?

Due to the several practical limitations associated with early software reliability predictions, this study focuses on studying fault prediction. A software system fails if any of the residual faults get executed, causing failure and making it unreliable. Although there may be a possibility that faults may exist in the code but never get executed for a certain input. This does not mean that it will not cause a failure in near future. If there are faults in the software, probability of failure also exists. Alternatively, unreliable software indicates that there are certainly some residual faults inside. Thus, unreliable software always means faulty software but the converse is not always true.

Realizing the importance of residual faults toward product reliability and quality, this book has primarily aims:

1. To predict the number of faults early in software life cycle using software metrics and process maturity.
2. To predict fault density indicators (FDIs) and residual faults in the software using a multistage model.
3. To develop an approach for prediction and ranking of fault-prone software module before testing using software quality metrics.
4. To propose a reliability centric test case prioritization method that will help the tester to detect faults cost-effectively during regression testing.
5. To develop operational profile of software-based product, computing the required number of test cases, generating test cases, and allocating test cases.

1.4 Software Fault Prediction

When errors occur during the software development process, faults or defects are introduced in the software. These faults are responsible for software failure or software unreliability/unavailability. Truly speaking, it is not possible to develop fault-free software in practical scenario considering human nature. Therefore, to ensure and control the reliability and quality of software, the faults present in the software must be known so that a mitigation actions can be planned accordingly. Knowing the faults early during software development helps software managers, to optimally allocate resources and achieve more reliable software within the time and cost constraints.

Generally, software reliability can be estimated or predicted using various available software reliability models (Musa et al. 1987; Pham 2006) based on failure data collected during testing. In case of early software reliability prediction, it is not possible due to the unavailability of failure data during early phase of software life cycle. Therefore, early reliability prediction can be assisted by early fault prediction by utilizing various available software metrics and development process maturity. Since the failure data are not available in the early phases of software life cycle, alternate information such as reliability relevant software metrics, developer's maturity level, and expert opinions can be utilized. This early fault information can help the project managers to reduce the software development cost by reducing the amount of the rework, fault finding, and debugging. Hence, early fault prediction seems to be a practical approach to ensure the reliability and quality of software system in the absence of failure data.

1.4.1 Software Metrics

IEEE (1988) had developed a standard IEEE STD 982.2 known as "IEEE Guide for the Use of IEEE Standard Dictionary of Measures to Produce Reliable Software." This guide provides the conceptual insights, implementation considerations, and suggestions for producing reliable software. The goal of the IEEE dictionary is to support software developers, project managers, and system users in achieving optimum reliability levels in software products.

1.4.2 Capability Maturity Model Level

The capability maturity model (CMM) has become a popular and widely accepted methodology to develop high-quality software within budget and time (Krishnan and Kellner 1999). For example, as one software unit at Motorola improved from CMM level 2 to 5, the average defect density reduced from 890 defects per million assembly equivalent lines of code to about 126 defects per million assembly equivalent lines (Diaz and Sligo 1997). In an empirical study on 33 software products developed by an IT company in 12 years, Harter et al. (2000) found that 1% improvement in process maturity resulted in 1.589% increase in product quality. In another study, Krishnan and Kellner (1999) have shown that process maturity and personnel capability to be a significant predictors (both at the 10% level) of the number of defects.

1.4.3 Limitation of Early Reliability Prediction Models

Early prediction of software reliability is useful for all stakeholders as it provides an opportunity for the early identification of faults which cause the failures. It also provides an insight toward optimal strategies toward software developments. One of the earliest and well-known efforts to predict software reliability in the earlier phase of the life cycle was the work initiated by the Rome Laboratory (1992). For their model, they developed prediction of fault density which they could then transform into other reliability measures such as failure rates. Recently, Kumar and Misra (2008) have proposed a model using software metrics. The major limitation with these models is that they have not discussed the way how to find software failure intensity function without executing the software, which is required to calculate the software reliability. Also, time parameter of reliability evaluation is not considered during reliability prediction. Reliability predictions are meaningless if time is not taken into considerations.

Due to the practical limitations associated with early software reliability prediction, this study is focused on early fault prediction models keeping software development process in consideration to improve the performance of early software reliability prediction. This early fault prediction lets the various stakeholder of software development to take appropriate action to mitigate residual faults that may cause failures in the future.

1.4.4 Early Software Fault Prediction Model

This work proposes a comprehensive framework to identify various reliability relevant software metrics applicable to early phase of software development. Then, fuzzy profiles and rules are developed using expert opinion. Finally, integrating these profiles and rules with fuzzy inference system (FIS), the number of faults at the end of each phase is obtained. The proposed model considers two most significant factors, software metrics, and process maturity together, for fault prediction. For this, it has collected various reliability relevant software metrics (IEEE 1988; Li and Smidts 2003) applicable to different phase of software development along with process maturity.

The model inputs are various reliability relevant software metrics and process maturity. The output of the model is the number of faults at the end of requirement phase, design phase, and coding phase. The basic steps of the model are identification of input/output variables, development of fuzzy profile of these input/output variables, defining relationships between inputs and output variables, and fault prediction at the end of each phase of software life cycle. This model helps software professionals to improve their development process and produce reliable software with minimum number of residual faults.

1.5 Residual Fault Prediction Model

Prediction of faults across the various phases of software development is desirable for any industry. It attracts both engineer and managers. Fault density seems to be a more desirable measure than the number of faults for reliability and quality indicator. Fault density metric can be found for requirement document, design document, code document, and also test report. After finding the fault density of test report, the number of faults present in the software can be predicted. The reliability of the software system can be measured using the number of residual faults that are likely to be found during testing or operational usage (Fenton et al. 2008). This work proposes a multistage fault prediction model which predicts the number of residual faults present in the software using software metrics. The model considers all development phase from requirement to testing and assume that the software is being developed through waterfall process model (Pressman 2005).

Stages of the proposed multistage model correspond to the phases of software development phases in one to one manner. Therefore, model is divided into four consecutive phases, that is, requirement, design, coding, and testing phase, respectively. Phase-I predicts the FDIs at the end of the requirement phase using relevant requirement metrics. Phase-II predicts the FDI at the end of the design phase using design metrics, and the fault density of requirements phase. Similarly, phase-III predict the FDI at the end of the coding phase using coding metrics and the fault density of design phase. Finally, the phase-IV predicts the FDI using testing metrics as well as FDI of coding phase. Finally, the FDI of testing phase is used for predicting the number of residual faults that are likely to be found during testing or operational usage.

1.5.1 Software Development Life Cycle and Fault Density

Software reliability has roots in each step of the software development life cycle and can be improved by inspection and review of these steps. Generally, the faults are introduced in each phase of software life cycle and keep on propagating to the subsequent phases unless they are detected through testing or review process.

Finally, undetected and/or uncorrected faults are delivered with software, causing failures. In order to achieve high software reliability, the number of residual faults in delivered code should be at minimum level.

Faults can be requirements faults, design faults, coding faults, and testing faults. These faults can be predicted throughout the development phase using different software metrics. For example, requirement faults can be predicted, using requirements metrics. Fault density measures have been widely used for software reliability assessments. Fault density (some time also referred as defect density) is a measure of the total known faults divided by the size of the software entity being measured and can be calculated as:

$$Fault\_density = \frac{No\_of\_fault}{size}$$

Size is typically counted either in lines of code or function points (FPs). For example, faults per KLOC or per function point are frequently used measuring code fault density. There are widespread belief that FPs is a better size metric than KLOC as they are language independent (Fenton and Neil 1999). Function point can be obtained from documents such as requirements, design, coding. Fenton and Neil (1999) provide defects density per function point for various documents as shown in Table 1.1.

In both cases, size normalizes comparisons between different software documents. Residual fault density can predicted at the end of a particular phase during software development life cycle. Examples include requirement fault density, design fault density, code fault density, and test fault density.

Fault density is a more appropriate measure of software reliability and quality as compared to the number of faults. We can justify this statement as—suppose software is being developed and we are at requirement analysis stage at some point in time, initially it is found that there is no requirement fault in the SRS (software requirements and specification) document. As time passes by and requirements get refined, developers start realizing that some faults are present in SRS document and this realization may grow till the completion of the phase. At the end of the phase, requirements fault density can be obtained by dividing the total number of requirements faults by size of requirements document. Therefore, it is intuitively clear that requirement fault density is a function of time. Similarly, design fault density, code fault density, and test fault density are function of time. Hence, fault density seems to be more suitable for reliability and quality prediction than the number of faults.

Table 1.1 Defects per function point

Defect origins	Defects per function point
Requirements	1.00
Design	1.25
Coding	1.75
Documentation	0.60
Bad fixes	0.40

1.5.2 Software Metrics and Fault Density Indicator

Fault density of a particular phase of software life cycle can be accurately calculated only if there is exact information about the number of faults as well as the size of that document is available. At the end of each phase, size is known but the number of exact residual faults cannot be estimated until it is exhaustively executed or implemented. Software metrics plays a vital role to assess or predict fault density, and so residual faults.

The term Fault Density Indicator (FDI) is used to represent fault density at the end of each phase of SDLC (Pandey and Goyal 2010). For early phases, an SRS document may use the metric such as requirements complexity (RC), requirements stability (RS), and review, inspection and walkthrough (RIW). Also size of the requirements document (SRS document) can be measured using different terms such as the number of requirements (functional and non-functional), function points. An SRS is a developer's understanding (in writing) of a customer or potential client's system requirements prior to any actual design or development work. It is an agreement document between developer and client at a given point in time. Considering variation in metrics and size estimates across the SDLC, FDI can be predicted. On the basis of FDI at the end of the testing phase, total number of residual faults in the software can be predicted as discussed in Chap. 4.

1.6 Quality of Large Software System

Predicting exact number of faults in software (or software modules) is often not necessary and also sometimes not desirable by the project manager where test efforts need to be prioritized and allocated to modules. Therefore, a need for classification of software module as fault-prone (FP) or not fault-prone (NFP) is realized.

Large software systems are developed by integrating various independent software modules. To ensure the quality and reliability of the system, it is important to assure the quality of various modules. Therefore, modules are tested independently and faults are removed as soon as they appear. The main problem with module testing is that testing time and resources are limited. Also, all the modules are neither equally important nor do they contain an equal amount of faults. Therefore, researchers started focusing on the classification of these modules as FP and NFP. The goal is to help software managers in prioritizing their testing efforts by focusing on those modules that are likely to contain more faults. Furthermore, by ranking FP modules on the basis of its degree of fault-proneness, testing effort can be optimized to achieve cost-effectiveness.

1.6.1 Fault-Prone and Not Fault-Prone Software Modules

The quality of a module is judged on the basis of the number of faults lying dormant in the module. A fault is a defect in the source code that causes failure when executed. A software module is said to be FP if the number of expected fault is more than a given threshold value. This threshold value is chosen according to project-specific criteria and can vary from project to project. In some cases, a module is said to be FP if there is a probability of finding at least one fault.

$$\text{Module} = \begin{cases} \text{FP,} & \text{if no.\_of\_faults} \geq \text{threshold} \\ \text{NFP,} & \text{if no.\_of\_faults} < \text{threshold} \end{cases}$$

A software module is stated to FP when there is a high probability of finding faults during its operation. Alternatively, the fault-proneness gives the value as how much probability is there that a module will contain a fault. Once modules are classified as FP or NFP, a cost-effective software quality improvement can be implemented to achieve reliable software within specified resource constraints.

1.6.2 Fault-Proneness Factors

Fault-proneness factors can be broadly classified as: product, process, and human factors. The following are some major factors that greatly affect the fault generation during the development of software modules.

Product factors include the structural or functional characteristic of modules such as size, cyclomatic complexity, and amount of coupling present among the modules. Process factors characterize the process followed at various phase of software development life cycle. These may include the amount of change in requirements document, design document, source code, and test cases. Besides, coding style (such as top-down or bottom-up), skills and experience of development team member also contribute a lot toward fault generation in the module. Human factors (Furuyama and Lio 1997) are concerned with the resources such as personnel, time, and cost assigned for the project under development. Nowadays, software is being developed by putting too much stress on the programmers. This may cause high rate of fault injection in the module. Therefore, the effects of stress on fault generation must be considered as an important human factors resulting fault-proneness of a software module. These stresses may be of different types such as indirect stress, direct stress, physical stress, and mental stress.

1.6.3 Need for Software Module Classification

Software quality prediction model is of a great interest among the software quality researchers and industry professionals. Software quality prediction can be done by predicting the number of software faults expected in the modules (Khoshgoftaar and Allen 1999; Khoshgoftaar and Seliya 2002) or classifying the software module as FP or NFP. On reviewing literature, it is found that supervised, semi-supervised, and unsupervised learning approaches have been used for building a fault prediction models. Among these, supervised learning approach is widely used and found to be more useful if sufficient amount of fault data from previous releases are available. In this case, software metrics and fault data from earlier or similar project are used in order to train the classification model, that is, to derive classifiers. After then, the derived classifiers are used for class prediction as FP and NFP.

1.6.4 Limitations with Earlier Module Prediction Models

The major problem with the traditional module prediction model is that they all are using the crisp values of software metrics and classify the module as a FP or NFP. It has been found that early phase software metrics have fuzziness in nature and crisp value assignment seems to be impractical. Also, all the traditional models are predicting the module as either FP or NFP. This type of prediction may suffer from some ambiguity and not desirable, where testing resources are to be allocated on the basis of its degree of fault-proneness. It is unwise to allocate equal amount of testing resource to all FP or NFP modules. A software module cannot be 100% FP or NFP. Some degree of fault-proneness is associated with each module. Therefore, FP modules can be ranked on the basis of its degree of fault-proneness. This enables the managers to allocate testing resources accordingly to achieve quality software in cost-effective manner.

Therefore, there is a need to classify module as FP and NFP, and at the same time, the FP modules are to be ranked on the basis of its degree of fault-proneness. This book presents a model for prediction and ranking for FP module using ID3 algorithm and FIS discussed in Chap. 5.

1.7 Regression Testing and Software Reliability

A software system requires various changes, throughout their lifetime, due to some residual faults, changes of user requirements, changes of environments, and so forth. It is important to ensure that these changes are incorporated properly without any adverse impact on the quality and reliability of the software. In general, when software are modified to correct faults or to add new function, new faults may be introduced. Therefore, software need to be retested after subsequent changes, in the form of regression testing. The objective of regression testing is to provide a general assurance that no additional errors were introduced in the process of fixing. To facilitate regression testing, suite of test cases is typically developed and stored with the program. It is very common to use these test suites to test a modified program by executing these test suites and find some regression faults, if any. Moreover, it is not only impractical but costly also to reexecute every test case for every program function if some changes occur. This makes regression testing very expensive among all testing process.

To reduce the cost of regression testing, researches have proposed many techniques such as regression test selection (RTS), test suite minimization (TSM), and test case prioritization (TCP). Both RTS and TSR techniques reduce the cost of regression by compromising the fault detection capability and have to sacrify the quality and reliability of the system under test. TCP techniques reduce the cost of regression testing without affecting the fault detection capabilities. The effectiveness of a TCP techniques was measured by the test suite's fault detection rate—a measure of how quickly a prioritize test suite detect faults. Problem with traditional TCP techniques is that they are not cost-effective, in particular, when number of test cases becomes more. Also these traditional TCP techniques do not consider the historical performance of the earlier test suite during prioritization. A cost-effective test case prioritization technique to improve the reliability of a software system is discussed in the Chap. 6 of this book.

1.8 Software Reliability and Operational Profile

System reliability (either hardware or software) depends on how it is being used in the field. The reliability of software system is strongly tied with its operational usages. It is a software fault that potentially causes failure when executed under particular conditions. A software fault may lead to system failure only if that fault is encountered during operational usage. If a fault is not accessed in a specific operational mode, it will not cause failures at all. Moreover, it will cause failure more often if a fault resides inside the frequently used code. Therefore, in software reliability engineering, it is good first to test most frequent code. Alternatively, prioritize testing of the software system or modules according to their profile.

The logic of testing with an operational profile attracts both software engineers and project managers because it identifies failures in the order of how often they occur and reliability information can be mapped in genuine way. This approach rapidly reduces failure intensity as test proceeds, and the faults that cause frequent failures are found and removed first. When we want to assure the reliability of a modified system that are already in use; operational profile–based testing is the best choice. Testing in accordance with an operational profile is therefore a key part of the test automation strategy. Reliability centric operational profile–based testing approach is presented in the Chap. 7 of this book.

1.9 Organization of the Book

Software faults are the root causes for software failures when get executed. These affect the reliability and quality of the software system. This book aim at developing fault prediction models toward software reliability and quality improvement. These developed models are applicable to the software development process and will be useful for both developer as well as end user. The book is divided into six chapters and three appendices as follows:

Chapter 2: Background: Software Quality and Reliability Prediction. This chapter describes the recent developments and the research work carried out in the field of software quality reliability prediction models, fault prediction models, software quality prediction models, regression testing, and operational profile.

Chapter 3: Early Fault Prediction Model Using Software Metrics Process Maturity. In this chapter, an early fault prediction model is proposed. This work is aimed to study fault prediction with a common goal of reliability and quality assurance of software systems. At the first, proposed model provides predicted faults during the early phases of software life cycle considering software metrics and process maturity. It is obvious that earlier a fault is identified, the better and more cost-effectively, it can be fixed. Another goal of the early fault prediction is to help developers to produce software with minimum number of residual faults through controlling most sensitive quality indices.

Chapter 4: Multistage Model for Residual Fault Prediction. After predicting the number of faults across the early phases of software development, a multistage model is proposed which predicts the number of residual faults that are likely to be found during testing or operational usage. Therefore, this chapter has proposed a multistage fault prediction model for residual fault prediction. This model considers all the software development phases and predicts the number of residual faults sitting dormant inside the software. Model predicts FDIs at the end of each phase of software development using relevant software metrics and FIS which is generic in nature. On the basis of FDIs at the end of the testing phase, the number of residual faults is predicted using a conversion equation.

Chapter 5: Prediction and Ranking of FP Software Module. In this chapter, an approach is proposed for prediction and ranking of FP software modules. Data mining techniques and software metrics available before testing are used for classification of FP software modules as well as its degree of fault-proneness. The goal is to help software manager to prioritize their testing efforts by focusing on those modules that are likely to contain faults.

Chapter 6: Reliability Centric Test Case Prioritization. This chapter has discussed an integrated and cost-effective approach to test prioritization that increases the test suite's fault detection rate. The proposed approach considers the three important factors, program change level (PCL), test suite change level (TCL), and test suite size (TS), before applying any techniques to execute test cases. These factors can be derived by using the information from the modified program version. A cost-effective reliability centric test case prioritization is also discussed in this chapter.

Chapter 7: Software Reliability and Operational Profile. The reliability of a software-based product is based on how the computer and other external elements will use it. Making a good reliability estimate of software depends on testing the product as per its field usages. If developed early, an operational profile may be used to prioritize the development process, so that more resources are put on the most important operations. This chapter discusses the way to develop the operational profile, computing the number of test case, and test case allocation. A case study using the traditional and operational profile–based testing is also discussed.

Appendix A: describes the metrics taken from IEEE standard 982.2 and Li and Smidts (2003) with their fuzzy profiles.

Appendix B: describes about "qqdefects" dataset which is publically available availed at the PROMISE repository dataset.

Appendix C: describes about KC2 dataset which is a NASA dataset and publically available at the PROMISE repository.

References

IEEE. (1991). *IEEE standard glossary of software engineering terminology.* STD-729-991, ANSI/IEEE.

Musa, J. D., Iannino, A., & Okumoto, K. (1987). *Software reliability: Measurement, prediction, and application.* McGraw–Hill Publication.

Lyu, M. R. (1996). *Handbook of software reliability engineering.* NY: McGraw–Hill/IEE Computer Society Press.

Pham, H. (2006). *System software reliability, reliability engineering series.* London: Springer.

Jelinski, Z., & Moranda, P. B. (1972). Software reliability research. In W. Freiberger (Ed.), *Statistical computer performance evaluation* (pp. 465–484). NY: Academic Press.

Shooman, M. L. (1972). Probabilistic models for software reliability prediction. In W. Freiberger (Ed.), *Statistical computer performance evaluation* (pp. 485–502). NY: Academic Press.

Littlewood, B., & Verrall, J. (1973). A bayesian reliability growth model for computer software. *Journal of the Royal Statistical Society, series C, 22*(3), 332–346.

Musa, J. D. (1975). A theory of software reliability and its application. *IEEE Transaction on Software Engineering, SE-1,* 312–327.

Schick, G. J., & Wolverton, R. W. (1978). An analysis of competing software reliability model. *IEEE Transaction on Software Engineering, SE-4*(2), 104–120.

Goel, A. L., & Okumoto, K. (1979). A time-dependent error detection rate model for software reliability and other performance measure. *IEEE Transaction on Reliability, R-28,* 206–211.

Kapur, P. K., & Garg, R. B. (1990). A software reliability growth model under imperfect debugging. *RAIRO, 24,* 295–305.

Chatterjee, S., Misra, R. B., & Alam, S. S. (1997). Joint effect of test effort and learning factor on software reliability and optimal release policy. *International Journal of System Science, 28*(4), 391–396.

Gaffney, G. E., & Pietrolewiez, J. (1990). An automated model for software early error prediction (SWEEP). In *Proceeding of 13th Minnow Brook Workshop on Software Reliability.*

Agresti, W. W., & Evanco, W. M. (1992). Projecting software defect from analyzing Ada design. *IEEE Transaction on Software Engineering, 18*(11), 988–997.

Rome Laboratory. (1992). *Methodology for software reliability prediction and assessment* (vol. 1–2). Technical Report RL-TR-92-52.

Yamada, S., Ohba, M., & Osaki, S. (1983). *S*-shaped reliability growth modelling for software error detection. *IEEE Transaction on Reliability, R-32,* 475–478.

IEEE. (1988). IEEE guide for the use of IEEE standard dictionary of measures to produce reliable software. *IEEE Standard 982.2.*

Krishnan, M. S., & Kellner, M. I. (1999). Measuring process consistency: implications reducing software defects. *IEEE Transaction on Software Engineering, 25*(6), 800–815.

Diaz, M., & Sligo, J. (1997). How software process improvement helped Motorola. *IEEE Software, 14*(5), 75–81.

Harter, D. E., Krishnan, M. S., & Slaughter, S. A. (2000). Effects of process maturity on quality, cycle time and effort in software product development. *Management Science, 46,* 451–466.

Kumar, K. S., & Misra, R. B. (2008). An enhanced model for early software reliability prediction using software engineering metrics. In *Proceedings of 2nd International Conference on Secure System Integration and Reliability Improvement* (pp. 177–178).

Li, M., & Smidts, C. (2003). A ranking of software engineering measures based on expert opinion. *IEEE Transaction on Software Engineering, 29*(9), 811–824.

Fenton, N., Neil, N., Marsh, W., Hearty, P., Radlinski, L., & Krause, P. (2008). On the effectiveness of early life cycle defect prediction with Bayesian Nets. *Empirical of Software Engineering, 13,* 499–537.

Pressman, R. S. (2005), *Software engineering: A practitioner's approach* (6th ed.). New York: McGraw-Hill Publication.

Fenton, N. E., & Neil, M. (1999). A critique of software defect prediction models. *IEEE Transaction on Software Engineering, 25*(5), 675–689.

Pandey, A. K., & Goyal, N. K. (2010). Fault prediction model by fuzzy profile development of reliability relevant software metrics. *International Journal of Computer Applications, 11*(6), 34–41.

Furuyama, A. Y., & Lio, K. (1997). Analysis of fault generation caused by stress during software development. *The Journal of Systems and Software, 38,* 13–25.

Khoshgoftaar, T. M., & Allen, E. B. (1999). A comparative study of ordering and classification of fault-prone software modules. *Empirical Software Engineering, 4,* 159–186.

Khoshgoftaar, T. M., & Seliya, N. (2002). Tree-based software quality models for fault prediction. In *Proceedings of 8th International Software Metrics Symposium, Ottawa, Ontario, Canada* (pp. 203–214).

Chapter 2
Background: Software Quality and Reliability Prediction

2.1 Introduction

Size, complexity, and human dependency on software-based products have grown dramatically during past decades. Software developers are struggling to deliver reliable software with acceptable level of quality, within given budget and schedule. One measure of software quality and reliability is the number of residual faults. Therefore, researchers are focusing on the identification of the number of fault presents in the software or identification of program modules that are most likely to contain faults. A lot of models have been developed using various techniques. A common approach is followed for software reliability prediction utilizing failure data. Software reliability and quality prediction is highly desired by the stakeholders, developers, managers, and end users. Detecting software faults early during development will definitely improve the reliability and quality in cost-effective way.

In this chapter, a review of the recent available literature on software reliability and quality prediction is presented. The following sections cover the literature surveys on software reliability models, reliability-relevant software metrics, software capability maturity models, software defect prediction model, and software quality prediction models, regression testing and test case prioritization, and operational profile-based testing.

2.2 Software Reliability Models

A number of analytical models have been presented in literature to address the problem of software reliability measurement. These approaches are based mainly on the failure history of software and can be classified according to the nature of the failure process. The various software reliability models may be categorized as failure rate model, fault count model, software reliability growth models, etc. Details about these models can be found in Musa et al. (1987), Goel and Okumoto (1979), Pham (2006), Lyu (1996). These reliability prediction models attempt to

A. K. Pandey and N. K. Goyal, *Early Software Reliability Prediction*,
Studies in Fuzziness and Soft Computing 303, DOI: 10.1007/978-81-322-1176-1_2,
© Springer India 2013

predict the reliability of the software in the later stages of the life cycle (testing and beyond). However, the opportunity of controlling software development process for cost-effectiveness is missed.

Software reliability models usually refer to estimating the number of remaining errors in a partially debugged software. Tests performed on the software derive outcome as accepted, conditionally accepted, or rejected. Acceptance may be based on the number of errors found over a selected period of time, on the number of paths executed of the total number of paths available to be executed, or some other prearranged criterion. Variety of reliability models are now competing for the attention of the analyst. Software reliability models can be broadly categorized as suggested by Pham (2006):

- *Deterministic* used to study the number of distinct operators and operands as well as the machine instructions in the program. Two most common deterministic models are:

  - Halstead's software metric, based on unique no. of operators and operands.
  - McCabe's cyclomatic complexity metric, based on cyclomatic number $V(G)$.

- *Probabilistic* describes the failure occurrence and\or fault removal phenomenon of the testing process as probabilistic events with respect to time and\or testing effort. Some common probabilistic models include the following (Pham 2006):

  - Failure rate model (times between failure models).
  - Failure or fault count model (NHPP models).
  - Error or fault-seeding model.
  - Reliability growth model, etc.

2.2.1 Failure Rate Models

Failure rate models are one of the earliest classes of models proposed for software reliability assessment. The most common approach is to assume that the time between $(i - 1)$ and the ith failures follows a distribution whose parameters depend on the number of faults remaining in the program during this interval. Estimates of the parameters are obtained from the observed values of times between failures or failure count in intervals. The estimates of software reliability, mean time to next failure, etc., are then obtained from the fitted model. The failure models can also be distinguished according to the nature of the relationship between the successive failures rates by Lyu (1996): (1) deterministic relationship, which is the case for most failure rate models and, (2) stochastic relationship. Following are the list of some failure rate models (Goel 1985):

- Jelinski and Moranda Model.
- Schick and Wolverton Model.
- Goel and Okumoto Imperfect Debugging Model, etc.

2.2.2 Failure or Fault Count Models

The interest of this class of models is in the number of faults or failures in specified time intervals rather than in times between failures. The failure counts are assumed to follow a known stochastic process with a time-dependent discrete or continuous failure rate. Parameters of the failure rate can be estimated from the observed values of failure counts or from failure times. Estimates of software reliability, mean time to next failure, etc., can again be obtained from the relevant equations. The key models in this class are Goel (1985) as follows:

- Shooman exponential model.
- Musa execution time model.
- Goel–Okumoto non-homogeneous Poisson process model (G–O-NHPP).
- S-shape growth model.
- Discrete reliability growth model.
- Musa-Okumoto logarithmic Poisson execution model.
- Generalized NHPP, etc.

The variety of existing NHPP models can be classified according to the several different classification systems (Kapur et al. 1990), and one important thing in this categorization is that they are not mutually disjoint. Some of these models are as follows:

- Modeling under perfect debugging environment.
- Modeling the imperfect debugging and error generation phenomenon.
- Modeling with testing effort.
- Testing domain-dependent software reliability modeling.
- Modeling with respect to testing coverage.
- Modeling the severity of faults.
- Software reliability modeling for distributed software systems.
- Modeling fault detection and correction with time lag.
- Managing reliability in operational phase.
- Software reliability assessment using SDE model.
- Neural network–based software reliability modeling, etc.

2.2.3 Error or Fault-Seeding Models

The basic approach in this class of models is to "seed" known number of faults in the program which is assumed to have an unknown number of indigenous faults. The program is tested and observed for numbers of seeded and indigenous faults. From these, an estimate of the indigenous fault content of the program is obtained and used to assess software reliability and other relevant measures. Mills' error seeding model (Mills 1972) is one the most popular and basic model to estimate the number of error in a program by introducing seeded errors into the program.

They have estimated unknown number of inherent errors from debugging data which consists of inherent and seeded errors.

Lipow (1972) modified Mills' model by taking into consideration the probability of finding a fault, of either kind, in any test of the software. Cai (1998) modified Mills' model and estimated the number of faults remaining in the software. Tohma et al. (1991) proposed a model for estimating the number of faults initially resident in a program at the beginning of the test or debugging process based on the hypergeometric distribution.

2.2.4 Reliability Growth Models

A software reliability growth model is applicable during the testing phase of software development and quantifies the software reliability in terms of estimated number of software error remaining in software or estimated time intervals between software failures. Software reliability growth is defined as the phenomenon that the number of software errors remaining in the system decreases with the progress of testing. For software growth modeling and analysis, the calendar time or CPU time is often used as the unit of software error detection period. However, the appropriate unit of software error detection period is sometimes the number of test runs or the number of executed test cases. A software reliability growth model for such a case is called a *discrete software reliability growth model*.

Software reliability growth models have been grouped into two classes of models: concave and S-shaped. These two model types are shown in Fig. 2.1. The most important thing about both models is that they have the same asymptotic behavior, that is, the defect detection rate decreases as the number of defects detected (and repaired) increases, and the total number of defects detected approach a finite value asymptotically (Table 2.1).

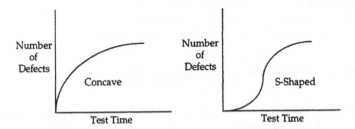

Fig. 2.1 *Concave* and *S-shaped* model

Table 2.1 Software reliability growth models (Wood 1996)

Model name	Model type	$\mu(t)$	Comments
Goel–Okumoto (G–O)	Concave	$a(1 - e^{-bt})$ $a \geq 0, b > 0$	Also called Musa model or exponential model
G–O S-shaped	S-shaped	$a(1 - (1 + bt)e^{-bt})$ $a \geq 0, b > 0$	Modification of G–O model to make it S-shaped
Hossain Dahiya	Concave	$a(1 - e^{-bt})/(1 + ce^{-bt})$ $a \geq 0, b > 0, c > 0$	Becomes same as G–O as c approaches to 0
Yamada exponential	Concave	$a(1 - e^{-r\alpha(1 - e^{-\beta t})})$ $a \geq 0, r\alpha > 0, \beta > 0$	Attempts to account for testing effort
Yamada Raleigh	S-shaped	$a\left(1 - e^{-r\alpha(1 - e^{(-\beta t^2/2)})}\right)$ $a \geq 0, r\alpha > 0, \beta > 0$	Attempts to account for testing effort
Weibull	Concave	$a(1 - e^{-bt^c})$ $a \geq 0, b > 0, c > 0$	Same as G–O for $c = 1$
Kapur et al.	Concave	$a(1 - e^{-bpt})$	Based on imperfect debugging

2.3 Architecture-based Software Reliability Models

The software reliability prediction models which are based on number of fault/
failures and times between failures are called black-box approach to predict
software reliability. The common feature of black-box models is the stochastic
modeling of the failure process, assuming some parametric model of cumulative
number of failures over a finite time interval or of the time between failures
(Gokhale et al. 2004). Variety of black-box reliability growth models can be found
in the books and papers such as Musa et al. (1987), Pham (2006). Lyu (1996), Goel
(1985).

The problem with these models is that the system is considered as a whole and
only its interaction with external world is modeled without considering its internal
architecture. Usually, in these models, no information is used except failure data
for assessing or predicting software reliability. Therefore, these black-box models
are not realistic to model a large component–based software system where various
software modules are developed independently. So, a need of white-box approach
is realized to estimate software reliability by considering the internal architecture
of software.

Thus, the main goal of architecture-based software reliability model is to
estimate the system reliability by considering the architecture of the software
components and their interaction with other ones. Also, these models assume that
components fail independently and that a component failure will ultimately lead to
a system failure. Failure can happen during an execution period of any module or
during the control transfer between two modules. The failure behavior of the
modules and of the interfaces between the modules can be specified in terms of
their reliabilities or failure rates. It is assumed that the transfer of control between

modules has a Markov property which means that given the knowledge of the module in control at any given time, the future behavior of the system is conditionally independent of the past behavior. Most of the architecture-based software reliability models are based on Markov process. A Markov process has the property that the future behavior of the process depends only on the current state and is independent of its past history. The various architecture-based softw are reliability model can be found in literatures such as Littlewood (1979), Popstojanova and Trivedi (2001), (2006), Gokhale et al. (2004), Gokhale (2007).

2.4 Bayesian Models

This group of models views reliability growth and prediction in a Bayesian framework rather than simply counting number of faults or failures. Failure and fault count model assumes that impact of each fault will be the same with respect to reliability. Also, these models allow change in the reliability only when failures occur. A Bayesian model takes a subjective viewpoint in that if no failures occur while the software is observed then the reliability should increase, reflecting the growing confidence in the software by the end user. The reliability is therefore a function of both the number of faults that have been detected and the amount of failure-free operation. This function is expressed in terms of a prior distribution representing the view from past data and a posterior distribution that incorporates past and current data.

The Bayesian models also reflect the belief that different faults have different impact on the reliability of the program. The number of faults is not as important as their impact. A program having number of faults in rarely used code will be more reliable than a program with only one fault in frequently used code. The Bayesian model says that it is more important to look at the behavior of the software than to estimate the number of faults in it. One of the very first models of this category is the Littlewood-Verrall reliability growth model (Littlewood and Verrall 1973).

2.5 Early Software Reliability Prediction Models

All the approaches discussed earlier for reliability prediction attempt to predict the reliability of the software in the later stages of the life cycle (testing and beyond). However, the opportunity of controlling software development process for cost-effectiveness is missed. Therefore, the need of early software reliability prediction is realized. Early reliability prediction attracts software professionals as it provides an opportunity for the early identification of software quality, cost overrun, and optimal development strategies. During the requirements, design, or coding phase, predicting the number of faults can lead to mitigating actions such as additional reviews and more extensive testing.

Gaffney and Pietrolewiez (1990) have developed a phase-based model to predict reliability during test and operation using fault statistics obtained during the technical review of requirements, design, and the coding phase. One of the earliest and well-known studies to predict software reliability in the earlier phase of the life cycle is the work initiated by the Rome laboratory (1992). For their model, they developed prediction of fault density which they could then be transformed into other reliability measures such as failure rates. Li et al. (2003) proposed a framework based on expert opinion elicitation, developed to select the software engineering measures which are the best software reliability indicators. In a similar direction, Kumar and Misra (2008) made an effort for early software reliability prediction considering the six top-ranked measures given by Li et al. (2003) and software operational profile.

2.6 Reliability-Relevant Software Metrics

In order to achieve high software reliability, the number of faults in delivered code should be reduced. Furthermore, to achieve the target software reliability efficiently and effectively, it needs to be known at early stages of software development process. One way of knowing software reliability during early stages of development is early software reliability prediction. Since early phase of software life cycle testing/field failure data is not available, information available such as reliability-relevant software metrics, developer's maturity level, and expert opinions can be utilized to predict the number of faults in the software.

IEEE had developed a standard IEEE Std. 982.2 (1988) known as "IEEE Guide for the Use of IEEE Standard Dictionary of Measures to Produce Reliable Software." The goal of the IEEE dictionary is to support software developers, project managers, and system users in achieving optimum reliability levels in software products. It was designed to address the needs of software developers and customers who are confronted with a surplus of models, techniques, and measures in the literature, but who lack sufficient guidance to utilize them effectively. The standard addresses the need for a uniform interpretation of these and other indicators of reliability. The IEEE dictionary assumes an intimate relationship between the reliability of a product and the process used to develop that product. The reliable product provides confirmation of a successful process; the unreliable product provides a lesson for process change. It is therefore the metrics selected for reliability assessment/prediction should provide insight into both process and product so that the essential facts necessary for process evaluation and required changes can be made effectively.

The measures are selected to provide information throughout the life cycle of a product. The basic goal is to provide the elements of a measurement program that support a constructive approach for achieving optimum reliability of the end product. The selected measures are related to various product and process factors that may have an impact on software reliability and that are observable throughout the life cycle. The IEEE dictionary provides, even in the early life cycle phases, a means for continual assessment of the evolving product and the process. Through early reliability assessment, the IEEE dictionary supports constructive optimization of reliability within given performance, resource, and schedule constraints.

The dictionary focuses on measures of the potential causes of failure throughout the software life cycle, rather than measures affecting reliability of nearly completed products. The intent is to produce software that is reliable, rather than just an estimate of reliability nearly completed and possibly unreliable product. Both the traditional approach of measuring reliability and the constructive approach of building in reliability are placed in context in this dictionary. The primitives to be used and the method of computation are provided for each measure. The standard calibrates the rulers, the measurement tools, through the use of common units and a common language. It promotes the use of a common database in the industry. Through commonality, the standard provides the justification for building effective tools for the measurement process itself. The standard and this guide are intended to serve as the foundation on which researchers and practitioners can build consistent methods. These documents are designed to assist management in directing product development and support toward specific reliability goals. The purpose is to provide a common set of definitions through which a meaningful exchange of data and evaluations can occur. Successful application of the measures is dependent on their use as intended in the specified environments.

Fenton (1991) has classified software metrics into three broad categories: product, process, and resources metrics. Product metrics describe characteristics of the product such as size, complexity, design features, performance, and quality level. Process metrics can be used to improve software development process and maintenance. Resources metrics describe the project characteristics and execution. Several researchers (e.g., Zhang and Pham 2000; Li et al. 2003) have shown that approximately thirty software metrics can be associated with different phases of software development life cycle. Among these metrics, some are significant predictor to reliability. Zhang and Pham (2000) have presented the findings of empirical research from 13 companies participating in software development to identify the factors that may impact software reliability. They have shown that thirty-two potential factors are involved during the various phases of software development process.

2.7 Software Capability Maturity Models

The capability maturity model (CMM) has become a popular methodology to develop high-quality software within budget and time. The CMM framework includes 18 key process areas such as quality assurance, configuration management, defect prevention, peer review, and training. A software process is assigned the highest maturity level if the practices in all 18 key process areas of the CMM are adopted. The CMM practices aid in reducing defect injection and in early identification of defects. As a consequence, the number of errors detected in testing and remaining in the delivered software will become lesser (Agrawal and Chari 2007). For example, as a software unit at Motorola improved from CMM level 2 to 5, the average defect density reduced from 890 defects per million assembly equivalent lines of code to about 126 defects per million assembly equivalent lines.

Paulk et al. (1993) provided an overview of the Capability Maturity Model for software development process. In their paper, they have discussed the software engineering and management practices that characterize organizations as they mature their processes for developing and maintaining software. Diaz et al. (1997) presented a case study that show average defect density reduced with increasing CMM level at Motorola. In a similar study, Krishnan and Kellner (1999) found that process maturity and personnel capability are significant predictors (both at the 10% level) of the number of defects. In an empirical study using 33 software products developed over 12 years by an IT company, Harter et al. (2000) found that 1% improvement in process maturity resulted in 1.589% increase in product quality. Agrawal et al. (2007) provided some results, which indicated that the biggest rewards from high levels of process maturity came from the reduction in variance of software development outcomes that were caused by factors other than software size.

2.8 Software Defects Prediction Models

Reliability of software system can be adversely affected by the number of residual faults present in the system. The main goal of software developers is to minimize the number of faults (defects) in the delivered code. An objective of all software projects is to minimize the number of residual faults in the delivered code and thus improving quality and reliability. Improving reliability is a key objective during system development and field deployment, and defect removal is the bottleneck in achieving this objective.

Lipow (1982) showed that the number of faults or "bugs" per line of code can be estimated based upon Halstead's software science relationships. This number is shown to be an increasing function of the number of lines of code in a program, a result in agreement with intuition, and some current theories of complexity. A similar kind of work has been carried out by Yu et al. (1988), in which they presented the results by analyzing several defect models using data collected from two large commercial projects.

In another study, Levendel (1990) proposed a birth–death mathematical model based on different defect behavior. Paper assumed that defect removal is ruled by the "laws of the physics" of defect behavior that controls the defect removal process. The time to defect detection, the defect repair time, and the factor of introduction of new defects due to imperfect defect repair are some of the "constants" in the laws governing defect removal. Test coverage is a measure of defect removal effectiveness. A model for projecting software defects from analyses of Ada design have been described by Agresti et al. (1992). The model predicted defect density based on the product and process characteristics. In a similar study, Wohlin et al. (1998) have presented two new methods to estimate the number of remaining defects after a review and hence control the software development process. The method is based on the review information from the individual reviewers and through statistical inferences. Conclusions about the remaining number of defects are then drawn after the reviews.

A critical review of software defect prediction model is provided by Fenton et al. (1999). This paper also proposes a model to improve the defect prediction situation by describing a prototype Bayesian belief network (BBN). A comprehensive evaluation of capture–recapture models for estimating software defect content was provided by Briand et al. (2000). Emam et al. (2001) have provided an extensive Monte Carlo simulation that evaluated capture–recapture models suitable for two inspectors assuming a code inspections context.

Fenton et al. (2008) presented a causal model for defect prediction using Bayesian nets. The main feature that distinguishes it from other defect prediction models is the fact that it explicitly combines both quantitative and qualitative factors. They have also presented a dataset for 31 software development projects. This dataset incorporates the set of quantitative and qualitative factors that were previously built into a causal model of the software process. The factors include values for code size, effort, and defects, together with qualitative data values judged by project managers using a questionnaire. Their model predicts the number of software defects that will be found in independent testing or operational usages with a satisfactorily predictive accuracy. Also, they have demonstrated that predictive accuracy increases with increasing project size. Catal et al. (2009) provided a systematic review of various software fault prediction studies with a specific focus on metrics, methods, and datasets.

Pandey and Goyal (2009) have developed an early fault prediction model using software metrics and process maturity which predicts the number of faults present before testing. The reliability of a software system depends on the number of residual faults sitting dormant inside. In fact, many of the software reliability models attempt to measure the number of residual bugs in the program (e.g., Briand et al. 2000; Emam et al. 2001; Fenton et al. 2008).

2.9 Software Quality Prediction Models

Assuring quality of large and complex software systems are challenging as these systems are developed by integrating various independent software modules. The quality of the system will depend on the quality of individual modules. All the modules are neither equally important nor do they contain an equal amount of faults. Therefore, researchers started focusing on the classification of these modules as fault-prone and not fault-prone.

Software reliability and quality prediction model is of a great interest among the software quality researchers and industry professionals. Software quality prediction can be done by predicting the expected number of software faults in the modules (Khoshgoftaar Allen 1999; Khoshgoftaar and Seliya 2002) or classifying the software module as fault-prone (FP) or not fault-prone (NFP). A commonly used software quality classification model is to classify the software modules into one of the following two categories: FP and NFP. A lot of efforts have been made for FP module prediction using various methods such as classification tree (Khoshgoftaar and Seliya 2002), neural networks (Singh et al. 2008), support vector machine (Singh et al. 2009) fuzzy logic (Kumar 2009) and logistic regression (Schneidewind 2001).

On reviewing literature, it is found that supervised (Menzies et al. 2007), semi-supervised (Seliya and Khoshgoftaar 2007a), and unsupervised learning (Catal and Diri 2008) approaches have been used for building a fault prediction models. Among these, supervised learning approach is widely used and found to be more useful FP module prediction if sufficient amount of fault data from previous releases are available. Generally, these models use software metrics of earlier software releases and fault data collected during testing phase. The supervised learning approaches cannot build powerful models with limited data. Therefore, some researchers presented a semi-supervised classification approach (Seliya and Khoshgoftaar 2007a) for software fault prediction with limited fault data. Unsupervised learning approaches such as clustering methods can be used in the absence of fault data. In most cases, software metrics and fault data obtained from a similar project or system release previously developed are used to train a software quality model. Subsequently, the model is applied to program modules of software currently under development for classifying them into the FP and NFP groups.

Various classification models have been developed for classifying a software module as FP and NFP. Schneidewind (2001) utilizes logistic regression in combination with Boolean discriminant functions for predicting FP software modules. Khoshgoftaar and Seliya (2002) incorporated three different regression tree algorithms CART-LS, S-PLUS, and CART-LAD into a single case study to show their effectiveness in finding the number of faults predicted using them. A study conducted by Khoshgoftaar and Seliya (2003) compared the fault prediction accuracies of six commonly used prediction modeling techniques. The study conducted a large-scale case study consisting of data collected over four successive system releases of a very large legacy telecommunications system. Some other works that have focused on FP module prediction include Munson and Khoshgoftaar (1992), Ohlsson and Alberg (1996), El-Emam et al. (2001).

Pandey and Goyal (2009) have presented an approach for prediction of the number of faults present in any software using software metrics. They have shown that software metrics are good indicators of software quality, and the number of faults present in the software. In certain scenarios, prediction of exact number of fault is not desirable and one needs to have the information about the quality of the software module. Software quality prediction can be done by predicting the number of software faults expected in the modules or classifying the software module as FP or NFP. Therefore, researchers started focusing on the classification of these modules as FP and NFP.

From the literature, it has been found that the decision tree induction algorithms such as CART, ID3, and C4.5 are efficient techniques for FP module classification. These algorithms uses crisp value of software metrics and classify the module as a FP or NFP. It has been found that early-phase software metrics have fuzziness in nature and crisp value assignment seems to be impractical. Also, a software module cannot be completely FP or NFP. In other words, it is unfair to assign a crisp value of software module representing its fault proneness.

2.10 Regression Testing

Regression testing is an important and expensive software maintenance activity to assure the quality and reliability of modified software. To reduce the cost of regression testing, researches have proposed many techniques such as regression test selection (RTS), test suite minimization (TSM), and test case prioritization (TCP). A test suite minimization technique is given by Harrold et al. (1993) to select a representative set of test cases from a test suite providing the same coverage as the entire test suite. Rothermel and Harrold (1996) have presented a framework for evaluating regression test selection techniques that classifies techniques in terms of inclusiveness, precision, efficiency, and generality. Wong et al. (1997) have found that both TSM and TCP suffers from certain drawbacks in some situation and suggested test case prioritization according to the criterion of increasing cost per additional coverage. Later, Rothermel et al. (1999) presented several techniques for prioritizing test cases and they empirically evaluated their ability to improve rate of fault detection—a measure of how quickly faults are detected within the testing process. For this, they provided a metric, APFD, which measures the average cumulative percentage of faults detected over the course of executing the test cases in a test suite in a given order. Their results have shown that test case prioritization can significantly improve the rate of fault detection of test suites. As a result of this, efforts have been made by many researchers on test case prioritization in order to improve fault detection rate (Elbaum et al. 2000, 2002, 2003, 2004; Do et al. 2006; Qu et al. 2007; Park et al. 2008; Khan et al. 2009; Kim and Baik 2010).

Review of literature indicate that earlier test case prioritization techniques have not considered the factors, such as program change level (PCL), test suite change level (TCL), and test suite size (TS) that affect the cost-effectiveness of the prioritization techniques. Also, all the traditional test case techniques presented to date have used a straightforward prioritization approach using some coverage criteria. We have found these traditional techniques are based on coverage information which relies on data gathered on the original version of a program (prior to modifications) in their prioritizations. They have ignored the information from the modified program version that may definitely affect the cost of prioritization.

2.11 Operational Profile

Reliability is a user-oriented view and strongly tied to the operational usage of the product. Operational profile becomes particularly valuable in guiding test planning by assigning test cases to different operations in accordance with their probabilities of occurrence. In the case of software, the operational usage information can be used to develop the various profiles such as customer profile, user profile, system mode profile, functional profile, and operational profile (Musa 2005).

One of the pioneer researches about the development of operational profile is by John D. Musa from AT&T Bell Laboratories (Musa 1993). It is a practical approach to ensure that a system is delivered with a maximum reliability, because the operations most frequently used also get tested the most. Musa informally characterized the benefits-to-cost ratio as 10 or greater (Musa 2005). In 1993, AT&T had used an operational profile successfully for the testing of a telephone switching service, which significantly reduced the number of problems reported by customers (Koziolek 2005). Also, Hewlett-Packard reorganized its test processes with operational profiles and reduced testing time and cost for a multiprocessor operating system by 50% (Koziolek 2005). Arora et al. (2005) conducted a case study on Pocket PC, a Windows CE 3.0-based device, and demonstrated a significant reliability improvement through operational profile–driven testing. It has been observed by many researchers and industries professional that operational profile–based testing is useful when the estimates of test cases are available (based on the constraints of the testing resource and time). Recently, Pandey et al. (2012) has presented a model-based approach to optimize the validation efforts by integrating the functional complexity and operational profile of fog light ECUs of an automotive system.

2.12 Observations

On reviewing literatures, the following observations have emerged:

1. Failure data are not available in the early phases of software life cycle and the information such as reliability-relevant software metrics, developer's maturity level, and expert opinions can be used. Both software metrics and process maturity play a vital role in early fault prediction in the absence of failure data.

Therefore, integrating software metrics with process maturity will provide a better fault prediction accuracy.

2. Early fault prediction is useful for both software engineers and managers since it provides vital information for making design and resource allocation decisions and thereby facilitates efficient and effective development process. Therefore, a model to predict number of faults present at the end of each phase of software life cycle is required. An early fault prediction model using software metrics process maturity is discussed in the Chap. 3.

3. One of the measures of software reliability is the number of residual faults and system reliability will be lesser as the number of residual defects (faults) in the system becomes more. There are several faults prediction models, but predicting faults without field failure data before testing are rarely discussed. Therefore, there is a need of a fault prediction model which predicts number of residual faults which may likely to occur after testing or operational use. Considering the importance of residual faults in reliability prediction, a multistage model for residual fault prediction is discussed in the Chap. 4.

4. Prediction of exact number of fault in a software module is not always necessary and there must be some measure of categorization which classify software module as FP or NFP. This will definitely help to improve the reliability and quality of software products by better resource utilization during software development process. Therefore, a model for prediction and ranking of FP software module is presented and discussed in the Chap. 5.

5. Regression testing is vital for reliability assurance of a modified program. It is one of the most expensive testings. Considering these points, a cost-effective reliability centric test case prioritization approach is parented in the Chap. 6.

6. The reliability of software, much more so than the reliability of hardware, is strongly tied to the operational usage of an application. Making a good reliability estimate of software depends on testing the product as per its field usages. Considering these points, reliability centric operational profile–based testing approach is discussed in the Chap. 7.

References

Musa, J. D., Iannino, A., & Okumoto, K. (1987). *Software reliability: measurement, prediction, and application*. New York: McGraw–Hill Publication.

Goel, A. L., & Okumoto, K. (1979). A time-dependent error detection rate model for software reliability and other performance measure. *IEEE Transaction on Reliability*, R-28, 206–211.

Pham, H. (2006). *System software reliability, reliability engineering series*. London: Springer.

Lyu, M. R. (1996). *Handbook of software reliability engineering*. NY: McGraw–Hill/IEE Computer Society Press.

Goel, A. L. (1985). Software reliability models: assumptions, limitations, and applicability. *IEEE Transaction on Software Engineering*, SE–11(12), 1411–1423.

Kapur, P. K., & Garg, R. B. (1990). A software reliability growth model under imperfect debugging. *RAIRO, 24,* 295–305.

Mills, H. D. (1972). *On the statistical validation of computer program* (pp. 72–6015). Gaithersburg, MD: IBM Federal Systems Division.

Lipow, M. (1972). *Estimation of software package residual errors.* Software Series Report TRW-SS-09, Redondo Beach, CA: TRW.

Cai, K. Y. (1998). On estimating the number of defects remaining in software. *Journal of System and Software,* 40(1).

Tohma, Y., Yamano, H., Ohba, M., & Jacoby, R. (1991). The estimation of parameter of the hypergeometric distribution and its application to the software reliability growth model. *IEEE Transaction on Software Engineering,* SE 17(2).

Wood, A. (1996). *Software reliability growth models.* Technical report 96.1, part number 130056.

Gokhale, S. S., Wong, W. E., Horgan, J. R., & Trivedi, K. S. (2004). An analytical approach to architecture-based software performance and reliability prediction. *Performance Evaluation, 58,* 391–412.

Littlewood, B. (1979). Software reliability model for modular program structure. *IEEE Transaction on Reliability,* R-28(3), 241–247.

Popstojanova, K. G., & Trivedi, K. S. (2001). Architecture-based approach to reliability assessment of software systems. *Performance Evaluation, 45,* 179–204.

Gokhale, S. S., & Trivedi, K. S. (2006). Analytical models for architecture-based software reliability prediction: a unification framework. *IEEE Transaction on Reliability, 55*(4), 578–590.

Gokhale, S. S. (2007). Architecture-based software reliability analysis: overview and limitations. *IEEE Transaction on Dependable and Secure Computing, 4*(1), 32–40.

Littlewood, B., & Verrall, J. (1973). A bayesian reliability growth model for computer software. *Journal of the Royal Statistical Society, series C, 22*(3), 332–346.

Gaffney, G. E., & Pietrolewiez, J. (1990). An automated model for software early error prediction (SWEEP). In *Proceeding of 13th Minnow Brook Workshop on Software Reliability.*

Rome Laboratory (1992). *Methodology for software reliability prediction and assessment* (Vols. 1–2). Technical report RL-TR-92-52.

Li, M., & Smidts, C. (2003). A ranking of software engineering measures based on expert opinion. *IEEE Transaction on Software Engineering, 29*(9), 24–811.

Kumar, K. S., & Misra, R. B. (2008). An enhanced model for early software reliability prediction using software engineering metrics. In *Proceedings of 2nd International Conference on Secure System Integration and Reliability Improvement* (pp. 177–178).

IEEE (1988). IEEE guide for the use of IEEE standard dictionary of measures to produce reliable software. *IEEE Standard 982.2.*

Fenton, N. (1991). *Software metrics: A rigorous approach.* London: Chapmann & Hall.

Zhang, X., & Pham, H. (2000). An analysis of factors affecting software reliability. *The Journal of Systems and Software, 50*(1), 43–56.

Agrawal, M., & Chari, K. (2007). Software effort, quality and cycle time: A study of CMM level 5 projects. *IEEE Transaction on Software Engineering, 33*(3), 145–156.

Paulk, M. C., Weber, C. V., Curtis, B., & Chrissis, M. B. (1993). Capability maturity model version 1.1. *IEEE Software, 10*(3), 18–27.

Diaz, M., & Sligo, J. (1997). How software process improvement helped Motorola. *IEEE Software, 14*(5), 75–81.

Krishnan, M. S., & Kellner, M. I. (1999). Measuring process consistency: implications reducing software defects. *IEEE Transaction on Software Engineering, 25*(6), 800–815.

Harter, D. E., Krishnan, M. S., & Slaughter, S. A. (2000). Effects of process maturity on quality, cycle time and effort in software product development. *Management Science, 46,* 451–466.

Lipow, M. (1982). Number of faults per line of code. *IEEE Transaction on Software Engineering,* SE–8(4), 437–439.

Yu, T. J., Shen, V. Y., & Dunsmore, H. E. (1988). An analysis of several software defect models. *IEEE Transaction on Software Engineering, 14*(9), 261–270.

Levendel, Y. (1990). Reliability analysis of large software systems: Defect data modeling. *IEEE Transaction on Software Engineering, 16*(2), 141–152.

Agresti, W. W., & Evanco, W. M. (1992). Projecting software defect from analyzing Ada design. *IEEE Transaction on Software Engineering, 18*(11), 988–997.

Wohlin, C. & Runeson, P. (1998). Defect content estimations from review data. In *Proceedings of 20th International Conference on Software Engineering* (pp. 400–409).

Fenton, N. E., & Neil, M. (1999). A critique of software defect prediction models. *IEEE Transaction on Software Engineering, 25*(5), 675–689.

Briand, L. C., Emam, K. E., Freimut, B. G., & Laitenberger, O. (2000). A comprehensive evaluation of capture: Recapture models for estimating software defect content. *IEEE Transaction on Software Engineering, 26*(8), 518–540.

El-Emam, K., Melo, W., & Machado, J. C. (2001). The prediction of faulty classes using object-oriented design metrics. *Journal of Systems and Software, 56*(1), 63–75.

Fenton, N., Neil, N., Marsh, W., Hearty, P., Radlinski, L., & Krause, P. (2008). On the effectiveness of early life cycle defect prediction with Bayesian Nets. *Empirical of Software Engineering, 13*, 499–537.

Catal, C., & Diri, B. (2009). Investigating the effect of dataset size, metrics set, and feature selection techniques on software fault prediction problem. *Information Sciences, 179*(8), 1040–1058.

Pandey, A. K., & Goyal, N. K. (2009). A fuzzy model for early software fault prediction using process maturity and software metrics. *International Journal of Electronics Engineering, 1*(2), 239–245.

Khoshgoftaar, T. M., & Allen, E. B. (1999). A comparative study of ordering and classification of fault-prone software modules. *Empirical Software Engineering, 4*, 159–186.

Khoshgoftaar, T. M., & Seliya, N. (2002). Tree-based software quality models for fault prediction. In *Proceedings of 8th International Software Metrics Symposium, Ottawa, Ontario, Canada* (203–214).

Khoshgoftaar, T. M., & Seliya, N. (2003). Fault prediction modeling for software quality estimation: comparing commonly used techniques. *Empirical Software Engineering, 8*, 255–283.

Singh, Y., Kaur, A., & Malhotra, R. (2008). *Predicting software fault proneness model using neural network*. LNBIP 9, Springer.

Singh, Y., Kaur, A., & Malhotra, R. (2009). Software fault proneness prediction using support vector machines. In *The Proceedings of the World Congress on Engineering, London, UK*, 1–3 July.

Kumar, K. S. (2009). Early software reliability and quality prediction (Ph.D. Thesis, IIT Kharagpur, Kharagpur, India).

Schneidewind, N. F. (2001). Investigation of logistic regression as a discriminant of software quality. In *The Proceedings of 7th International Software Metrics Symposium, London, UK* (pp. 328–337).

Menzies, T., Greenwald, J., & Frank, A. (2007). Data mining static code attributes to learn defect predictors. *IEEE Transactions on Software Engineering, 33*(1), 2–13.

Seliya, N., & Khoshgoftaar, T. M. (2007). Software quality estimation with limited fault data: A semi-supervised learning perspective. *Software Quality Journal, 15*(3), 327–344.

Catal, C., & Diri, B. (2008). A fault prediction model with limited fault data to improve test process. In *Proceedings of the 9th International Conference on Product Focused Software Process Improvement* (pp. 244–257).

Munson, J. C., & Khoshgoftaar, T. M. (1992). The detection of fault-prone programs. *IEEE Transactions on Software Engineering, 18*(5), 423–433.

Ohlsson, N., & Alberg, H. (1996). Predicting fault-prone software modules in telephone switches. *IEEE Transaction on Software Engineering, 22*(12), 886–894.

Harrold, M., Gupta, R., & Soffa, M. (1993). A methodology for controlling the size of a test suite. *ACM Transaction on Software Engineering and Methodology, 2*(3), 270–285.

Rothermel, G., & Harrold, M. J. (1996). Analyzing regression test selection techniques. *IEEE Transaction on Software Engineering, 22*(8), 529–551.

Wong, W. E., Horgan, J. R., London, S. & Agrawal, H. (1997). A study of effective regression testing in practice. In *Proceedings of the Eighth Int'l Symposium on Software Reliability Engineering* (pp. 230–238).

Rothermel, G., Untch, R. H., Chu, C., & Harrold, M. J. (1999). Test case prioritization: An empirical study. In *Procedings of the. Int'l Conf. Software Maintenance* (pp. 179–188).

Elbaum, S., Malishevsky, A., & Rothermel, G. (2000). Prioritizing test cases for regression testing. In *Proceedings of the International Symposium on Software Testing and Analysis* (pp. 102–112).

Elbaum, S., Malishevsky, A., & Rothermel, G. (2002). Test case prioritization: a family of empirical studies. *IEEE Transaction of Software Engineering, 28*(2), 159–182.

Elbaum, S., Kallakuri, P., Malishevsky, A., Rothermel, G., & Kanduri, S. (2003). Understanding the effects of changes on the cost-effectiveness of regression testing techniques. *Journal of Software, Verification and Reliability, 12*(2), 65–83.

Elbaum, S., Rothermel, G., Kanduri, S., & Malishevsky, A. G. (2004). Selecting a cost-effective test case prioritization technique. *Software Quality Journal, 12*(3), 185–210.

Do, H., Rothermel, G., & Kinneer, A. (2006). Prioritizing Junit test cases: An empirical assessment and cost-benefits analysis. *Empirical Software Engineering, 11*, 33–70.

Qu, B., Nie, C., Xu, B. & Zhang, X. (2007). Test case prioritization for black box testing. In *The Proceedings of 31st Annual International Computer Software and Applications Conference*.

Park, H., Ryu, H., & Baik, J. (2008). Historical value-based approach for cost-cognizant test case prioritization to improve the effectiveness of regression testing. In *The Proceedings 2nd International Conference on Secure System Integration and Reliability Improvement* (pp. 39–46).

Khan, S. R., Rehman, I., & Malik, S. (2009). The impact of test case reduction and prioritization on software testing effectiveness. In *Proceeding of International Conference on Emerging Technologies* (pp. 416–421).

Kim, S., & Baik J. (2010). An effective fault aware test case prioritization by incorporating a fault localization technique. In *Proceedings of ESEM-10, Bolzano-Bozen, Italy* (pp. 16–17).

Musa, J. D. (2005). *Software reliability engineering: more reliable software faster and cheaper* (2nd ed.). Tata McGraw-Hill Publication.

Musa, J. D. (1993). Operational profiles in software reliability engineering. *IEEE Software Magazine*.

Koziolek, H. (2005). Operational profiles for software reliability. Seminar on Dependability Engineering, Germany.

Arora, S., Misra, R. B., & Kumre, V. M. (2005). Software reliability improvement through operational profile driven testing. In *Proceedings of Annual IEEE Conference on Reliability and Maintainability Symposium, Virginia* (pp. 621–627).

Pandey, A. K., Smith, J., & Diwanji, V. (2012). Cost effective reliability centric validation model for automotive ECUs. In *The 23rd IEEE International Symposium on Software Reliability Engineering, Dallas, TX USA* (pp. 38–44).

Chapter 3
Early Fault Prediction Using Software Metrics and Process Maturity

3.1 Introduction

Development of reliable software is challenging as system engineers have to deal with a large number of conflicting requirements such as cost, time, reliability, safety, maintainability, and many more. These days, most of the software development tasks are performed in labor-intensive way. This may introduce various faults across the development, causing failures in the near future. The impact of these failures ranges from marginal to catastrophic consequences. Therefore, there is a growing need to ensure the reliability of these software systems as early as possible.

IEEE defines software reliability as "the probability of a software system or component to perform its intended function under the specified operating conditions over the specified period of time" (ANSI/IEEE 1991). In another way, this can also be defined as "the probability of failure-free software operation for a specified period of time in a specified environment." Software reliability is generally accepted as the key factor of software quality since it quantifies software failures, which make the system inoperative or risky (Agrawal and Chari 2007).

A software failure is defined as "the departure of external result of program operation from requirements," whereas a fault is defined as "the defect in the program when executed under particular conditions, causes a failure" (Musa et al. 1987). To further elaborate, a software fault is a defective, missing, or extra instruction or set of related instructions that is caused by one or more actual or potential failures.

Software reliability has roots in each step of the requirements, design, and coding processes (Kaner 2004) and can be improved by inspection and review of these steps. Also, this can be accurately assessed only after the testing or after the product completion. Generally, software reliability can be estimated or predicted using various available software reliability models (Musa et al. 1987; Pham 2006). These models use failure data collected during testing, and reliability can be estimated or predicted from a fitted model. This becomes too late and sometimes infeasible for taking corrective actions. The solution to this problem is to predict the software reliability in the early stage of development process, that is, before testing. This early reliability information can help in project management in

A. K. Pandey and N. K. Goyal, *Early Software Reliability Prediction,*
Studies in Fuzziness and Soft Computing 303, DOI: 10.1007/978-81-322-1176-1_3,
© Springer India 2013

reducing the development cost by reducing the amount of the rework. Moreover, since the failure data are not available during the early phases of software life cycle, the information such as process-level software metrics, expert opinions, and similar or earlier project data can be used for characterizing the factors affecting software reliability.

This chapter has focused on the identification of reliability-relevant software metrics for early fault prediction. For this, a comprehensive framework has been proposed to gather reliability-relevant software metrics from the early phase of software development life cycle, processing it, and integrating it with the fuzzy logic system to predict the number of faults present in the software at different phases before testing. This chapter is organized in the following way. Section 3.2 provides a brief overview of fuzzy logic system for early fault prediction. Section 3.3 discusses the proposed model. Section 3.4 discusses the model implementation in detail. Section 3.5 contains the case studies and results, whereas conclusions are presented in Sect. 3.6.

3.2 Brief Overview of Fuzzy Logic System

Fuzzy logic was originally introduced by Zadeh (1965), as a mathematical way to represent vagueness in everyday life. Fuzzy logic, as its name suggests, is the logic underlying modes of reasoning which are approximate rather than exact. The importance of fuzzy logic derives from the fact that most of the modes of human reasoning and especially commonsense reasoning are approximate in nature. Therefore, fuzzy logic inference systems have found usefulness in capturing and processing subjective information in terms of software metrics during the early phase of software development. Most of the early phases of software metrics are not very clear and involve high and complex dependency among them. Therefore, fuzzy logic is considered to be an appropriate tool in these situations.

Figure 3.1 shows an overview of fuzzy logic system for fault prediction. There are four main components of a fuzzy logic system, namely fuzzification process (input), fuzzy rule base, fuzzy inference process, and defuzzification (output). The fuzzification is the process of converting a crisp quantity into fuzzy form. In other words, fuzzification provides a mapping from crisp value to the fuzzy value. Depending on the data available and associated uncertainty, the input and output parameters are fuzzified in terms of linguistic variables such as low (L), medium (M), and high (H). The basic unit of any fuzzy system is the fuzzy rule base. All other components of the fuzzy logic system are used to implement these rules in a reasonable and efficient manner. The fuzzy inference process combines the rules in the fuzzy rule base and then provides a mapping from fuzzy input (input membership) to fuzzy output (output membership). There may be situations where the output of a fuzzy process needs to be single scalar quantity (crisp value) as opposed to a fuzzy set. To obtain a crisp value from fuzzy output, the fuzzy output is to be defuzzified. Defuzzification is the conversion of a fuzzy quantity to a

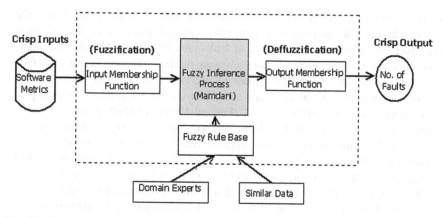

Fig. 3.1 Fuzzy logic system for fault prediction

precise quantity, just as fuzzification is the conversion of a precise quantity to a fuzzy quantity.

The output of a fuzzy process can be the logical union of two or more fuzzy membership functions defined on the universe of discourse of the output variable. For example, a fuzzy output comprises of two parts: a trapezoidal membership shape and a triangular membership shape. Union of these two membership functions that involves the max operator, which graphically is the outer envelope of the two shapes, is shown in the Fig. 3.2.

There are many defuzzification methods (Ross 2005) used in the literature in recent years such as max-membership method, centroid method, weighted average method, mean–max membership, middle of maximum, largest of maximum, and smallest of maximum. Centroid method is used in this study for the proposed model. The centroid method is discussed in the Sect. 3.4.3.

3.3 Proposed Model

Early fault prediction provides an opportunity for the early identification of software quality, cost overrun, and optimal development strategies. During the requirements, design or coding phases predicting the number of faults can lead to mitigating actions

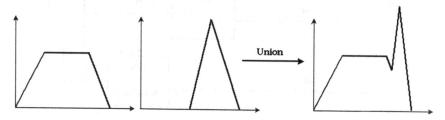

Fig. 3.2 A simple fuzzy output process

such as additional reviews and more extensive testing (Gaffney et al. 1990). The model is based on the assumption that the reliability and quality of a software system are adversely affected by the number of faults present in the software. The model considers two most significant factors, software metrics and process maturity (PM) together, for fault prediction. The model architecture is shown in Fig. 3.3.

Software metrics can be classified into three categories: product metrics, process metrics, and resources metrics (Fenton 1991). Product metrics describe characteristics of the product such as size, complexity, design features, performance, and quality level. Process metrics can be used to improve software development process and maintenance. Resources metrics describe the project characteristics and execution. Approximately thirty software metrics exist, which can be associated with different phases of software development life cycle (Zhang et al. 2000; Li et al. 2003). Among these metrics, some are significant predictor to reliability (Li et al. 2003).

Almost every industry is using the Capability Maturity Model (CMM) level to define their PM and improvements. One of the widely adopted frameworks is the CMM developed by the software engineering institute (SEI) at Carnegie Mellon University (Paulk et al. 1993). Based on the specific software practices adopted, the CMM classifies the software PM into five maturity levels, CMM level 1–5. Each level process maturity reduces defect density by a factor of 2, improving the overall quality and reliability (Diaz and Sligo 1997).

The CMM framework includes 18 key process areas such as quality assurance, configuration management, defect prevention, peer review, and training (Paulk et al. 1993). A software process is assigned the highest maturity level if the practices in all

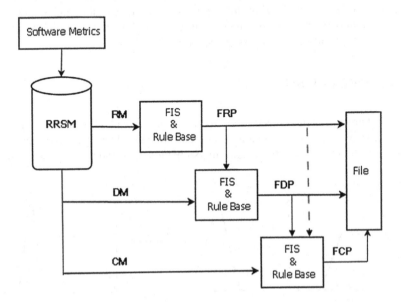

Fig. 3.3 Early fault prediction model

18 key process areas of the CMM are adopted. As per CMM methodology, an effective and mature software process must include interactions among employees, tools, and methods to accomplish any task. The CMM practices aid in reducing defect injection and in the early identification of defects. As a consequence, the number of errors detected in testing and remaining in the delivered software will become lesser (Diaz and Sligo 1997).

The proposed model gathers reliability-relevant software metrics (RRSMs) from various available sources (IEEE 1988; Li et al. 2003; NASA 2004; Kumar 2008). Table 3.1 shows the list of various metrics, which are considered as input/ output variable for the proposed model. The model considers three requirements' metrics (RM), *viz.* requirements' change request (RCR); review, inspection, and walk-through (RIW); and PM, as input to the requirements' phase. Similarly, at design phase, three design metrics (DM); design defect density (DDD); fault-days number (FDN), and data flow complexity (DFC), are considered as input. Two coding metrics (CM), code defect density (CDD) and cyclomatic complexity (CC), have been taken as input at coding phase. The outputs of the model will be the number of faults at the end of requirements' phase (FRP), number of faults at the end of design phase (FDP), and number of faults at the end of coding phase (FCP).

3.3.1 Description of Metrics Considered for the Model

3.3.1.1 Requirements' Change Request

It is necessary to understand requirements before the design and coding begins. Understanding the requirements of a problem is the most difficult task during software development. Changes in requirements are the fact and may arise at any time during the life cycle of a software development. Requirements' changes may occur to provide enhancement of the features of the system or to respond to competition, changing customer needs, and other factors. These changes may be for the adaptation of a system to the changes that occur either in hardware or in software. Further, requirements for a software project may change in order to provide necessary or desirable system performance improvements. Studies have shown that most of the software faults are traced back to ambiguous or ill-defined requirements. RCR measure provides insight into system reliability and quality. Uncontrolled changes in the requirements may lead to adverse effect on cost, quality, and schedule of project under development.

Table 3.1 Input/output variables

Phase	Input variables	Output variables
Requirement	RCR, RIW, PM	FRP
Design	FRP, DDD, FDN, DC	FDP
Coding	FRP, FDP, CDD, CC	FCP

In order to reduce the number of faults in the software, RCR should be minimized. More number of RCR may lead to more number of faults in the subsequent phases of software development. Qualitative range of RCR is gathered on a three-scale value viz, low (L), medium (M), and high (H), and fuzzy profiles are developed accordingly. It has been found that an RCR value up to 30 is considered as low by most of the people; similarly, 25–70 RCR is considered as medium and the value 60 and above is found to be high. Therefore, the range of RCR in the proposed model is considered from 1 to 100. The fuzzy profile of RCR is given in Table 3.3 and shown in Fig. 3.4.

3.3.1.2 Review, Inspection, and Walk-through

Software reviews are a "filter" for the software process and can be applied at various points during software development. Software reviews purify the software engineering activities, and there are many different types of reviews that can be conducted during software development. A formal presentation of software requirements and specification (SRS) document to the customers, management, and technical staff is called formal technical review, inspection, and walk-through (RIW). RIW is the most effective filter from a quality assurance standpoint. In software industries, the RIW is an effective means for uncovering errors and improving software quality.

Like RCR, the qualitative ranges of RIW value is gathered on a three-scale value, viz, low (L), medium (M), and high (H), and fuzzy profiles are developed accordingly. It is found that the value 0–1 of RIW is considered to be low for most of the people, the value 2–3 is considered as medium, and the value 3.5 is found to be sufficient for review and inspection. Therefore, the range of RIW in the proposed model may be considered from 1 to 5. The fuzzy profile of RIW is given in Table 3.3 and shown in Fig. 3.5.

3.3.1.3 Process Maturity

In many companies, the CMM plays a major role in defining software process improvement. CMM level is one of the widely accepted frameworks by the various

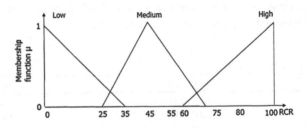

Fig. 3.4 Fuzzy profile of RCR

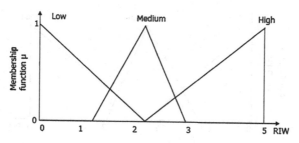

Fig. 3.5 Fuzzy profile of RIW

industries to define and improve their PM. Based on the specific software practices adopted, the CMM classifies the software PM into five maturity levels, CMM level 1–5. Each level PM reduces defect density by a factor of 2, improving the overall quality and reliability (Diaz and Sligo 1997). These improvements are thought to arise from reduced defects and rework in a mature software development process.

There are many experts who argue that the PM of the development process is the best predictor of product quality and hence of the residual faults (Fenton and Neil 1999). The simplest metric of process quality is the five-level ordinal-scale SEI CMM ranking. There are evidences that level $(n + 1)$ companies generally deliver products with lower residual defect density than level (n) companies (Diaz 1997; Fenton and Neil 1999; Harter et al. 2000; Agrawal and Chari 2007). Defect potentials and delivered defects for different CMM levels are shown in Table 3.2. This study has considered PM as SEI CMM level with five levels. Similar to the RCR and RIW, a value between 0 and 1 is considered as low PM; one to three is considered as medium maturity by most of the experts; and finally, a value 3 and above is considered as high maturity. The fuzzy profile of PM is given in Table 3.3 and shown in Fig. 3.6.

Table 3.2 CMM level and defect potential

SEI CMM levels	Defect potential	Delivered defects	Removal efficiency (%)
1	5	0.75	85
2	4	0.44	89
3	3	0.27	91
4	2	0.14	93
5	1	0.05	95

Table 3.3 Requirements' phase metrics

Value	RCR (0–100)	RIW (0–5)	PM (0–5)	FRP (0–100)
Very low				(0; 0; 5)
Low	(0; 0; 35)	(0; 2; 2)	(0; 2; 2)	(2; 5; 8)
Medium	(25; 45; 75)	(1; 2; 3)	(1; 2; 3)	(5; 10; 20)
High	(60; 100; 100)	(2; 5; 5)	(2; 5; 5)	(15; 30; 45)
Very high				(30; 100; 100)

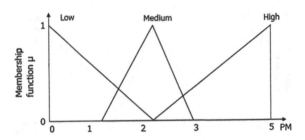

Fig. 3.6 Fuzzy profile of PM

3.3.1.4 Design Defect Density

A problem detected in the same phase where it was introduced is defined as an *error*; a *defect* is a problem that escapes detection in the phase it was introduced (Diaz 1997). A defect (also referred as fault) is nothing but a human mistake. DDD is calculated after inspection of new design document or old design documents after some modifications and can be calculated as

$$DDD = \frac{\sum_{i=1}^{n} DDi}{KSLOD}$$

where DD_i is the total number of defects found during ith design inspection, n is the number of inspection to date, and KSLOD is the number of source line in design documents in thousands.

These defects are found through the inspection of design documents once it has developed. The objective is to find the design defects, if any, which can result in more faults during its implementation. For this, developed design document is given to inspection team member to inspect the document independently and note down their findings. The DDD is considered as an important metric for early fault prediction (IEEE 1988). The fuzzy profile of DDD is given in Table 3.4 and shown in Fig. 3.7.

3.3.1.5 Fault-days Number

This measure represents the number of days that faults spend in the software system from their creation to their removal. The basic primitives are as follows: (1)

Table 3.4 Design phase metrics

Value	DDD (0–5)	FDN (0–50)	DFC (0–500)	FDP (0–100)
Very low				(0; 0; 10)
Low	(0; 0; 2)	(0; 0; 5)	(0; 0; 100)	(5; 10; 15)
Medium	(1; 2; 3)	(2; 7; 15)	(80; 160; 250)	(10; 20; 30)
High	(2; 5; 5)	(10; 50; 50)	(200; 500; 500)	(20; 40; 60)
Very high				(50; 100; 100)

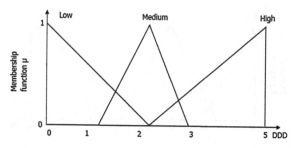

Fig. 3.7 Fuzzy profile of DDD

phase, when the fault was introduced in the system; (2) date, when the fault was introduced in the system; and (3) phase, date, and time, when the fault is removed.

For each fault detected and removed, during any phase, the number of days from its creation to its removal is determined (fault days). The fault days are then summed for all faults detected and removed, to get the FDN at system level, including all faults detected/removed up to the delivery date. In cases, when the creation date for the fault is not known, the fault is assumed to have been created at the middle of the phase in which it was introduced. Fault days for the design fault for module can be easily and accurately calculated because the design approval date for the detailed design of module is known.

The measure is calculated as the effectiveness of the software design and development process. It depends on the timely removal of faults across the entire life cycle. This measure is an indicator of the quality of the software system design and of the development process. For instance, a high number of fault days indicate either many faults (probably due to poor design, indicative of product quality) or long-lived faults or both. Therefore, FDN acts as good reliability-relevant software metric for early software fault prediction. The fuzzy profile of FDN is developed using Table 3.4 and shown in Fig. 3.8.

3.3.1.6 Data Flow Complexity

This is a structural complexity or procedural complexity measure that can be used to evaluate the following: (1) the information flow structure of large-scale systems;

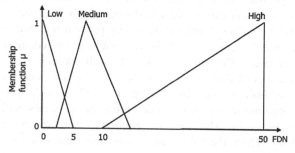

Fig. 3.8 Fuzzy profile of FDN

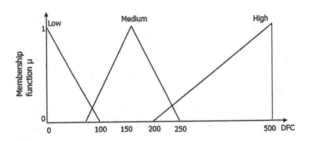

Fig. 3.9 Fuzzy profile of DFC

(2) the procedure and module information flow structure; and (3) the complexity of the interconnections between modules.

This measure is an indicator of the quality of the software system design. For instance, a module with a high information flow complexity is assumed to be likely to have more faults than a module with a low complexity value. This measure, generally, is applied first during detailed design and reapplied during integration to confirm the initial implementation. For large-scale systems, an automated procedure for determining data flow is essential. Moreover, this measure can also be used to indicate the degree of simplicity of relationships between subsystems and to correlate total observed failures and software reliability with data complexity. Therefore, DFC is considered as good reliability-relevant software metric for early software fault prediction. The fuzzy profile of DFC is developed using Table 3.4 and shown in Fig. 3.9.

3.3.1.7 Code Defect Density

CDD is calculated after code inspection of new development or large block modification and can be calculated as,

$$CDD = \frac{\sum_{i=1}^{n} CD_i}{KSLOC}$$

where CD_i is the total number of defects found during ith code inspection, n is the number of inspections to date, and KSLOC is the number of source lines of executable code and non-executable data declaration in thousands.

The defects are found throughout the coding phase using code inspection and walk-through. Once the coded module is successfully compiled, most of the syntax errors get eliminated. The objective of code walk-through is to discover the algorithmic and logical errors present in the code. For this, code is given to the review team member, a few days before the walk-through meeting. Review members are required to review the code, select the test cases, and simulate the code execution manually. Also, they are required to note the finding during these review processes. Then, the CDD can be calculated using the above equation.

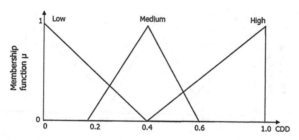

Fig. 3.10 Fuzzy profile of CDD

CDD is found to be an important metric for early fault prediction (IEEE 1988). The fuzzy profile of CDD is developed using Table 3.5 as shown in Fig. 3.10.

3.3.1.8 Cyclomatic Complexity

CC metric provides a quantitative measure of the logical complexity of a program. This measure is given by Pressman (2005) and is based on control flow graph. For a control flow graph with "N" nodes, "E" edges, and "P" connected components, the CC $V(G)$ is obtained using the following equation:

$$V(G) = E - N + P.$$

The condition and control statements may add or reduce the complexity of the program. It is generally intuitive that the program with the larger number of decision statements is likely to be more complex. CC, $V(G)$, for a flow graph G, can also be obtained as follows:

$$V(G) = D + 1.$$

The $V(G)$ value provides an upper bound for the number of independent paths. For any software project, the minimum number of CC will be of unity and maximum number of CC according to NASA Metric Data Program is 456 (NASA 2004). Therefore, the range of CC in the proposed model may be considered from 1 to 500. The range of CC with its fuzzy profile is given in Table 3.5 and shown in Fig. 3.11.

Table 3.5 Coding phase metrics

Value	CDD (0–1)	CC (0–500)	FCP (0–100)
Very low			(0; 0; 15)
Low	(0; 0; 0.4)	(0; 0; 150)	(5; 15; 30)
Medium	(0.2; 0.4; 0.6)	(100; 200; 300)	(15; 35; 60)
High	(0.4; 1; 1)	(250; 500; 500)	(30; 50; 80)
Very high			(60; 100; 100)

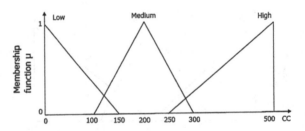

Fig. 3.11 Fuzzy profile of CC

3.4 Implementation of the Proposed Model

The model is implemented in MATLAB utilizing fuzzy logic toolbox. The basic steps of the model are identification of reliability-relevant software metrics (input/output variables), development of fuzzy profile of these input/output variables, defining relationships between input and output variables, and fault prediction at the end of each phase of software life cycle using fuzzy inference system (FIS). These basic steps can be grouped into four broad phases as follows: (1) early information gathering phase, (2) information processing phase, (3) defuzzification, and (4) fault prediction phase.

3.4.1 Information Gathering Phase

The quality of the fuzzy approximation depends mainly on the quality of information collected (subjective knowledge) and expert opinion. The information gathering phase is often considered the most vital step in developing a fuzzy logic system. The development of fuzzy profiles of input/output variables and fuzzy rules is assumed as building blocks of the fuzzy inference system and includes three major steps as discussed below.

3.4.1.1 Identification of Input/Output Variables

Tables 3.3, 3.4, and 3.5 list the various reliability-relevant software metrics with their fuzzy profile. These metrics are taken from IEEE (1988), Li et al. (2003), and their fuzzy profiles are developed using the documents such as IEEE (1988), NASA (2004), and Kumar (2009). A total of eight top-ranked reliability-relevant metrics have been identified. These metrics are considered as input variables and can be applied to the various software life cycle phases for fault prediction. Apart from that, three output variables such as FRP, FDP, and FCP are also taken. FRP, FDP, and FCP represent the number of faults at the end of requirements, design, and coding phases, respectively. These input/output variables with the corresponding fuzzy range and numbers are given in Tables 3.3, 3.4, and 3.5. Input variables are the reliability-relevant software metrics, and output variables are the number of faults at the end of each phase. In order to justify the selection of these software metrics in the proposed model, regression analysis has been performed for different metrics and corresponding correlation coefficient has been identified, besides other factors such as time, ease, and cost in finding these software metrics.

3.4.1.2 Development of Fuzzy Profiles

This is the first step in incorporating human knowledge into engineering systems in a systematic and efficient manner. The data, which may be useful for selecting appropriate linguistic variable, are generally available in one or more forms of (Kumar et al. 2008):

- Expert's opinion.
- Software requirements.
- User's expectations.
- Record of existing field data from previous release or similar system.
- Marketing professional's opinion.
- Data obtained from system logs, etc.

Input/output variables gathered at the previous steps are fuzzy in nature and are characterized by membership function. Triangular membership functions are considered for fuzzy profile development of various identified input/output variables. Triangular membership functions (TMFs) are widely used for calculating and interpreting reliability data because they are simple and easy to understand (Yadav et al. 2003). Also, they have the advantage of simplicity and are commonly used in reliability analysis (Ross 2005). Fuzzy membership functions are generated utilizing the linguistic categories such as very low (VL), low (L), medium (M), high (H), and very high (VH), identified by a human expert to express his/her assessment. Fuzzy profiles of all the various considered input/output variables are shown in Figs. 3.4, 3.5, 3.6, 3.7, 3.8, 3.9, 3.10, 3.11, 3.12, 3.13, and 3.14 for visualization purpose.

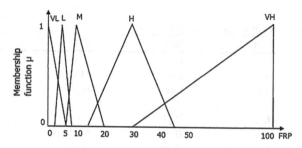

Fig. 3.12 Fuzzy profile of FRP

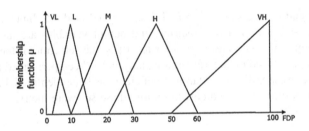

Fig. 3.13 Fuzzy profile of FDP

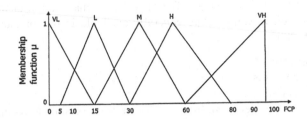

Fig. 3.14 Fuzzy profile of FCP

3.4.1.3 Development of Fuzzy Rule Base

The most important part of the early fault prediction system is the rules and how they interact with each other in order to generate results. The rules come from the experts so that the expert system can emulate the inference of an actual expert. To develop fuzzy rule base, knowledge can be acquired from different sources such as domain experts, historical data analysis of similar or earlier system, and engineering knowledge from existing literature's (Xie et al. 1999; Zhang and Pham 2000). In the present study, rules are generated from the software engineering point of view, and all of them take the form of "If A then B." Tables 3.6, 3.7, and 3.8 show the fuzzy rules used in the FIS for fault prediction.

Table 3.6 Fuzzy rules at requirements' phase

Rule	RCR	RIW	PM	FRP
1	Low	Low	Low	Very low
2	Low	Low	Medium	Low
3	Low	Low	High	Medium
⋮	⋮	⋮	⋮	⋮
25	High	High	Low	Medium
26	High	High	Medium	Very high
27	High	High	High	Very high

Table 3.7 Fuzzy rules at design phase

Rule	FRP	DDD	FDN	DFC	FDP
1	Very low	Low	Low	Low	Very low
2	Very low	Low	Low	Medium	Low
3	Very low	Low	Low	High	Medium
⋮	⋮	⋮	⋮	⋮	⋮
133	Very high	High	High	High	High
134	Very high	High	High	High	Very high
135	Very high	High	High	High	Very high

Table 3.8 Fuzzy rules at coding phase

Rule	FRP	FDP	CC	CDD	FCP
1	Very low	Very low	Low	Low	Very low
2	Very low	Very low	Low	Medium	Very low
3	Very low	Very low	Low	High	Low
⋮	⋮	⋮	⋮	⋮	⋮
223	Very high	Very high	High	Low	High
224	Very high	Very high	High	Medium	Very high
225	Very high	Very high	High	High	Very high

3.4.2 Information Processing Phase

In this phase, the fuzzy system maps all the inputs to an output. This process of mapping inputs onto output is known as fuzzy inference process or fuzzy reasoning (Zadeh 1989; Bowles and Pelaez 1995). Basis for this mapping is the number of fuzzy IF–THEN rules, each of which describes the local behavior of the mapping. The two main activities for information processing are as follows: combining input from all the "if" part fuzzy rules and aggregation of "then" part to produce the final output. The Mamdani fuzzy inference system (Mamdani 1977) is considered here for all the information processing.

3.4.3 Defuzzification

Defuzzification is the process of deriving a crisp value from a fuzzy set using any defuzzification methods such as centroid, bisector, middle of maximum, largest of maximum, and smallest of maximum (Ross 2005). Centroid method is used in the present study for finding the crisp value, representing the number of faults at the end of each phase of software life cycle. Centroid method is also called centre of area or centre of gravity method. This method is the most prevalent and physically appealing of all defuzzification methods (Ross 2005). The centroid value can be calculated as follows:

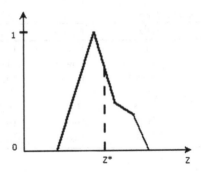

Fig. 3.15 Centroid defuzzification

$$Z = \frac{\sum_{i=1}^{N} \mu i \, di}{\sum_{i=1}^{N} \mu i},$$

if the values are discrete, where z is the crisp value and d_i is the value from the set that have a membership value μ_i.

$$Z = \frac{\int \mu(z) \cdot z \, dz}{\int \mu(z) \, dz},$$

if the values are continuous, where $\int$ denotes an algebraic integration. This method is shown in Fig. 3.15.

3.4.4 Fault Prediction Phase

Stages present in the proposed structure are shown in Fig. 3.1. The model resembles waterfall model (Pressman 2005). It divides the structure into three consecutive phases I, II, and III, which are, requirements' phase, design phase, and coding phase, respectively. Phase-I predicts the number of faults at the end of requirements' phase using requirement metrics such as RCR, RIW, and PM. Phase-II predicts the number of faults at the end of design phase using design metrics such as DDD, FDN, and DFC. Since most of the software faults traced back to requirements error, FRP is considered as input metric to this phase. Similarly, at phase-III, besides the coding metrics (CC and CDD), FRP and FDP are also considered as input to predict the number of faults at the end of design phase. In this study, a max–min Mamdani inference process was used for early fault prediction. The inference process uses the rules given in Tables 3.6, 3.7, and 3.8, and centroid method is used for defuzzification.

Table 3.9 Results at requirements' phase for project-I

	RCR	RIW	PM	FRP
Worst case	100.0	0.0	0.0	70.0
Average case	50.0	2.5	2.5	10.0
Best case	0.0	5.0	5.0	1.3

Table 3.10 Results at design phase for project-I

	DDD	FDN	DFC	FDP
Worst case	5.0	50.0	500.0	81.5
Average case	2.5	25.0	250.0	40.0
Best case	0.0	0.0	0.0	3.2

Table 3.11 Results at coding phase for project-I

	CDD	CC	FCP
Worst case	1.0	500.0	83.0
Average case	0.5	250.0	53.6
Best case	0.0	0.0	6.1

Table 3.12 Results at requirements' phase for project-II

	RCR	RIW	PM	FRP
Worst case	100.0	0.0	0.0	66.0
Average case	50.0	2.5	2.5	8.0
Best case	0.0	5.0	5.0	2.4

Table 3.13 Results at design phase for project-II

	DDD	FDN	DFC	FDP
Worst case	5.0	50.0	500.0	69.8
Average case	2.5	25.0	250.0	39.0
Best case	0.0	0.0	0.0	2.6

Table 3.14 Results at coding phase for project-II

	CDD	CC	FCP
Worst case	1.0	500.0	81.9
Average case	0.5	250.0	51.2
Best case	0.0	0.0	4.3

3.5 Case Studies

In order to analyze the fault prediction consistency and effect of various software metrics on early fault prediction, the values of the metrics from four different software projects are considered.

Table 3.15 Results at requirements' phase for project-III

	RCR	RIW	PM	FRP
Worst case	100.0	0.0	0.0	62.0
Average case	50.0	2.5	2.5	10.0
Best case	0.0	5.0	5.0	1.6

Table 3.16 Results at design phase for project-III

	DDD	FDN	DFC	FDP
Worst case	5.0	50.0	500.0	72.0
Average case	2.5	25.0	250.0	41.1
Best case	0.0	0.0	0.0	5.0

Table 3.17 Results at coding phase for project-III

	CDD	CC	FCP
Worst case	1.0	500.0	77.5
Average case	0.5	250.0	46.7
Best case	0.0	0.0	5.5

Table 3.18 Results at requirements' phase for project-IV

	RCR	RIW	PM	FRP
Worst case	100.0	0.0	0.0	77.0
Average case	50.0	2.5	2.5	15.0
Best case	0.0	5.0	5.0	5.5

Table 3.19 Results at design phase for project-IV

	DDD	FDN	DFC	FDP
Worst case	5.0	50.0	500.0	80.8
Average case	2.5	25.0	250.0	45.0
Best case	0.0	0.0	0.0	6.3

Table 3.20 Results at coding phase for project-IV

	CDD	CC	FCP
Worst case	1.0	500.0	83.7
Average case	0.5	250.0	50.7
Best case	0.0	0.0	6.6

The number of faults at the end of requirements, design and coding phases for four different projects is shown in the Tables 3.9, 3.7, 3.8, 3.9, 3.10, 3.11, 3.12, 3.13, 3.14, 3.15, 3.16, 3.17, 3.18, 3.19, and 3.20. Best-case and worst-case input metrics are applied to the proposed model in order to find the lower and upper bounds of prediction range at each phases. For example, in case of project 1, the number of faults at the end of requirements' phase may range from 1.3 to 70.0; the number of faults at the end of design phase may range from 3.2 to 81.5; the number of faults at the end of coding phase may range from 6.1 to 83.0.

Similarly, for project 2, the number of faults at the end of requirements' phase may range from 2.4 to 66.0; the number of faults at the end of design phase may range from 2.6 to 69.8; the number of faults at the end of coding phase may range from 4.3 to 81.9. For project 3, the number of faults at the end of requirements' phase may range from 1.6 to 62.0; the number of faults at the end of design phase may range from 5.0 to 72.0; the number of faults at the end of coding phase may range from 5.5 to 77.5. Finally, for project 4, the number of faults at the end of requirements' phase may range from 5.5 to 77.0; the number of faults at the end of design phase may range from 6.3 to 80.8; the number of faults at the end of coding phase may range from 6.6 to 83.7.

Table 3.21 Sensitivity of FRP with respect to change in RCR

RIW = 0 % PM = 0 %		RIW = 25 % PM = 25 %		RIW = 50 % PM = 50 %		RIW = 75 % PM = 75 %		RIW = 100 % PM = 100 %	
RCR	FRP	RCR	FRP	RCR	FRP	RCR	FRP	RCR	FRP
0.0	1.3	0.0	8.4	0.0	10.0	0.0	1.5	0.0	1.3
10.0	1.3	10.0	8.4	10.0	10.0	10.0	1.5	10.0	1.3
20.0	1.3	20.0	8.4	20.0	10.0	20.0	1.5	20.0	1.3
30.0	21.6	30.0	8.4	30.0	10.0	30.0	3.7	30.0	3.4
40.0	25.0	40.0	10.0	40.0	10.0	40.0	5.0	40.0	5.0
50.0	25.0	50.0	10.0	50.0	10.0	50.0	5.0	50.0	5.0
60.0	25.0	60.0	10.0	60.0	10.0	60.0	5.0	60.0	5.0
70.0	61.3	70.0	22.2	70.0	23.3	70.0	8.9	70.0	8.9
80.0	70.0	80.0	25.0	80.0	25.0	80.0	10.0	80.0	10.0
90.0	70.0	90.0	25.0	90.0	25.0	90.0	10.0	90.0	10.0
100.0	70.0	100.0	25.0	100.0	25.0	100.0	10.0	100.0	10.0

Table 3.22 Sensitivity of FRP with respect to change in PM

RCR = 0 % RIW = 0 %		RCR = 25 % RIW = 25 %		RCR = 50 % RIW = 50 %		RCR = 75 % RIW = 75 %		RCR = 100 % RIW = 100 %	
PM	FRP	PM	FRP	PM	FRP	PM	FRP	PM	FRP
0.0	1.3	0.0	2.5	0.0	25.0	0.0	25.0	0.0	25.0
1.0	50.0	1.0	50.0	1.0	50.0	1.0	50.0	1.0	50.0
2.0	5.0	2.0	7.3	2.0	10.0	2.0	10.0	2.0	10.0
3.0	5.0	3.0	7.6	3.0	10.0	3.0	10.0	3.0	10.0
4.0	10.0	4.0	10.0	4.0	10.0	4.0	10.0	4.0	10.0
5.0	10.0	5.0	10.0	5.0	10.0	5.0	10.0	5.0	10.0

Table 3.23 Sensitivity of FRP with respect to change in RIW

RCR = 0 % PM = 0 %		RCR = 25 % PM = 25 %		RCR = 50 % PM = 50 %		RCR = 75 % PM = 75 %		RCR = 100 % PM = 100 %	
RIW	FRP	RIW	FRP	RIW	FRP	RIW	FRP	RIW	FRP
0.0	1.3	0.0	5.0	0.0	10.0	0.0	10.0	0.0	10.0
1.0	1.3	1.0	5.0	1.0	10.0	1.0	10.0	1.0	10.0
2.0	5.0	2.0	10.0	2.0	10.0	2.0	10.0	2.0	10.0
3.0	5.0	3.0	10.0	3.0	10.0	3.0	10.0	3.0	10.0
4.0	5.0	4.0	1.9	4.0	10.0	4.0	10.0	4.0	10.0
5.0	5.0	5.0	1.9	5.0	10.0	5.0	10.0	5.0	10.0

3.6 Results and Discussion

The proposed model is applied to four different software projects, and the faults are predicted by utilizing the varying values of corresponding metrics across the different phases. The model helps to determine the impact of a particular software metrics on the software faults. Once the impact of the particular software metrics on fault has been identified, the better and more cost effectively it can be controlled to improve the overall reliability and quality of the product. For each phase, a sensitivity analysis is performed to find the impact of various software metrics on the number of faults.

Table 3.24 Sensitivity of FDP with respect to change in DDD

FDN = 0 % DFC = 0 %		FDN = 25 % DFC = 25 %		FDN = 50 % DFC = 50 %		FDN = 75 % DFC = 75 %		FDN = 100 % DFC = 100 %	
DDD	FDP	DDD	FDP	DDD	FDP	DDD	FDP	DDD	FDP
0.0	3.0	0.0	10.8	0.0	25.0	0.0	25.0	0.0	25.0
1.0	3.0	1.0	10.8	1.0	25.0	1.0	25.0	1.0	25.0
2.0	3.0	2.0	25.0	2.0	45.0	2.0	45.0	2.0	45.0
3.0	8.6	3.0	35.6	3.0	45.0	3.0	45.0	3.0	45.0
4.0	10.7	4.0	35.6	4.0	45.0	4.0	45.0	4.0	45.0
5.0	10.7	5.0	35.6	5.0	45.0	5.0	45.0	5.0	45.0

Table 3.25 Sensitivity of FDP with respect to change in FDN

DDD = 0 % DFC = 0 %		DDD = 25 % DFC = 25 %		DDD = 50 % DFC = 50 %		DDD = 75 % DFC = 75 %		DDD = 100 % DFC = 100 %	
FDN	FDP	FDN	FDP	FDN	FDP	FDN	FDP	FDN	FDP
0.0	3.0	0.0	10.7	0.0	25.0	0.0	25.0	0.0	25.0
10.0	3.3	10.0	17.3	10.0	25.0	10.0	25.0	10.0	25.0
20.0	10.7	20.0	17.2	20.0	45.0	20.0	45.0	20.0	45.0
30.0	10.7	30.0	17.2	30.0	45.0	30.0	45.0	30.0	45.0
40.0	10.7	40.0	17.2	40.0	45.0	40.0	45.0	40.0	45.0
50.0	10.7	50.0	17.2	50.0	45.0	50.0	45.0	50.0	45.0

Table 3.26 Sensitivity of FDP with respect to change in DFC

DDD = 0 % FDN = 0 %		DDD = 25 % FDN = 25 %		DDD = 50 % FDN = 50 %		DDD = 75 % FDN = 75 %		DDD = 100 % FDN = 100 %	
DFC	FDP	DFC	FDP	DFC	FDP	DFC	FDP	DFC	FDP
0.0	3.0	0.0	17.6	0.0	25.0	0.0	25.0	0.0	25.0
100.0	10.8	100.0	19.3	100.0	25.0	100.0	45.0	100.0	45.0
200.0	10.7	200.0	19.3	200.0	25.0	200.0	45.0	200.0	45.0
300.0	25.0	300.0	35.6	300.0	45.0	300.0	45.0	300.0	45.0
400.0	25.0	400.0	35.6	400.0	45.0	400.0	45.0	400.0	45.0
500.0	25.0	500.0	35.6	500.0	45.0	500.0	45.0	500.0	45.0

From Tables 3.21, 3.22, and 3.23, it has been found that at the requirements' phase, RCR metric has more impact on the number of faults as compared to RIW and PM. From Table 3.21, it was shown that for a constant value of RIW and PM, increment in the value of RCR results in the number of requirements' faults. When RCR value is small, that is, less than 20, FRP value is found to be very less, that is,

Table 3.27 Sensitivity of FCP with respect to change in CC

CDD = 0 %		CDD = 25 %		CDD = 50 %		CDD = 75 %		CDD = 100 %	
CC	FCP	CC	FCP	CC	FCP	CC	FCP	CC	FCP
0.0	5.7	0.0	11.8	0.0	13.9	0.0	20.0	0.0	20.0
100.0	5.7	100.0	11.8	100.0	13.9	100.0	20.0	100.0	20.0
200.0	5.7	200.0	11.8	200.0	20.0	200.0	40.0	200.0	40.0
300.0	20.0	300.0	20.0	300.0	50.0	300.0	40.0	300.0	40.0
400.0	20.0	400.0	20.0	400.0	50.0	400.0	40.0	400.0	40.0
500.0	20.0	500.0	20.0	500.0	50.0	500.0	40.0	500.0	40.0

Table 3.28 Sensitivity of FCP with respect to change in CDD

CC = 0 %		CC = 25 %		CC = 50 %		CC = 75 %		CC = 100 %	
CDD	FCP	CDD	FCP	CDD	FCP	CDD	FCP	CDD	FCP
0.0	5.67	0.0	6.4	0.0	5.67	0.0	20.0	0.0	20.0
0.2	5.67	0.2	6.4	0.2	5.67	0.2	20.0	0.2	20.0
0.4	13.9	0.4	15.4	0.4	20.0	0.4	50.0	0.4	50.0
0.6	15.4	0.6	27.3	0.6	31.9	0.6	40.0	0.6	40.0
0.8	20.0	0.8	31.9	0.8	40.0	0.8	40.0	0.8	40.0
1.0	20.0	1.0	31.9	1.0	40.0	1.0	40.0	1.0	40.0

Table 3.29 Comparison of prediction results

	Requirements' phase			Design phase			Coding phase		
	Worst case	Average case	Best case	Worst case	Average case	Best case	Worst case	Average case	Best case
Kumar (2009)	70.0	2.0	1.0	79.0	45.0	3.0	85.0	67.0	6.0
Proposed model	70.0	11.3	1.3	81.5	40.0	3.2	83.0	53.6	6.1

1.3. However, for higher value of RCR, after a particular threshold, there is a sudden increase in the value of FRP. This threshold value is found to be 30 for RCR. Similarly, sensitivity of FRP with respect to change in PM can be found from Tables 3.22–3.23, which gives the sensitivity of FRP with respect to change in RIW.

Similarly, at the design phase, DDD and DFC are found to have more influence on the number of fault as compared to FDN. Tables 3.24, 3.25, and 3.26 give the sensitivity of FDP with respect to change in DDD, FDN, and DFC, respectively. At the coding phase, both CDD and CC have the approximate equal effect on the number of faults, when different values are applied to the effect on the number of faults.

Tables 3.27 and 3.28 give the sensitivity of FCP with respect to change in CC and CDD, respectively. It would be helpful for project managers to know the importance of various software metrics across the software development phase and and effect of these metrics on sofware faults for cost-effective management.

The prediction results are compared with the earlier work (Kumar 2009) and are shown in Table 3.29. The prediction results are found promising when maximum, minimum, and average values of the software metrics are applied to the model. This model can be useful for software developer to find the minimum, maximum, and average number of faults exiting in their project during the development. This fault prediction is expected to provide early indication of software reliability and quality and to help software managers to optimize efforts. For this, the developers have to provide the values of the various matrices of the phase for which they are intended to find the number of faults.

3.7 Summary

A model for the early prediction of software fault is presented in this chapter. The model proposed in this chapter has utilized software metrics and PM to predict the number of faults present at the end of the early phase of software life cycle. A total of eight reliability-relevant metrics have identified, and using fuzzy inference system, a total number of faults at the end of the early phases of software life cycle are predicted. The prediction consistency and validity are verified through for different projects. For software professionals, this model provides an insight into software metrics and its impact on software fault during the development process. For software project mangers, the model provides a methodology for allocating the resources for developing reliable and cost-effective software. The model will be useful in reducing the number of residual faults in the delivered software.

References

IEEE (1991). IEEE standard glossary of software engineering terminology. ANSI/IEEE, STD-729–991.

Agrawal, M., & Chari, K. (2007). Software effort, quality and cycle time: A study of CMM level 5 projects. *IEEE Transaction on Software Engineering, 33*(3), 145–156.

Musa, J. D., Iannino, A., & Okumoto, K. (1987). *Software reliability: Measurement, prediction, and application.* New York: McGraw–Hill Publication.

Kaner, C. (2004). Software engineering metrics: What do they measure and how do we know? In *10th International Software Metrics Symposium, METRICS.*

Pham, H. (2006). *System software reliability, reliability engineering series.* London: Springer.

Zadeh, L. A. (1965). *Fuzzy sets, information and control* (Vol. 8(3) pp. 338–353).

Ross, T. J. (2005). *Fuzzy logic with engineering applications* (2nd ed.). India: Willey.

Gaffney, G. E., & Pietrolewiez, J. (1990). An automated model for software early error prediction (SWEEP). In *Proceeding of 13th Minnow Brook Workshop on Software Reliability.*

Fenton, N. (1991). *Software metrics: A rigorous approach.* London: Chapmann & Hall.

Zhang, X., & Pham, H. (2000). An analysis of factors affecting software reliability. *The Journal of Systems and Software, 50*(1), 43–56.

Li, M., & Smidts, C. (2003). A ranking of software engineering measures based on expert opinion. *IEEE Transaction on Software Engineering, 29*(9), 811–824.

Paulk, M. C., Weber, C. V., Curtis, B., & Chrissis, M. B. (1993). Capability maturity model version 1.1. *IEEE Software, 10*(3), 18–27.

Diaz, M., & Sligo, J. (1997). How software process improvement helped Motorola. *IEEE Software, 14*(5), 75–81.

IEEE (1988). IEEE guide for the use of ieee standard dictionary of measures to produce reliable software. *IEEE Standard 982.2.*

NASA (2004). NASA metrics data program. http://mdp.ivv.nasa.gov/.

Kumar, K. S., & Misra, R. B. (2008). An enhanced model for early software reliability prediction using software engineering metrics. In *Proceedings of 2nd International Conference on Secure System Integration and Reliability Improvement* (pp. 177–178).

Fenton, N. E., & Neil, M. (1999). A critique of software defect prediction models. *IEEE Transaction on Software Engineering, 25*(5), 675–689.

Harter, D. E., Krishnan, M. S., & Slaughter, S. A. (2000). Effects of process maturity on quality, cycle time and effort in software product development. *Management Science, 46*, 451–466.

Pressman, R. S. (2005). *Software engineering: A practitioner's approach* (6th ed.). New York: McGraw-Hill Publication.

Yadav, O. P., Singh, N., Chinnam, R. B., & Goel, P. S. (2003). A fuzzy logic based approach to reliability improvement during product development. *Reliability Engineering and System Safety, 80*, 63–74.

Xie, M., Hong, G. Y., & Wohlin, C. (1999). Software reliability prediction incorporating information from a similar project. *The Journal of Systems and Software, 49*, 43–48.

Zadeh, L. A. (1989). Knowledge representation in fuzzy logic. *IEEE Transactions on Knowledge and Data Engineering, 1*, 89–100.

Bowles, J. B., & Pelaez, C. E. (1995). Application of fuzzy logic to reliability engineering. *Proceedings of IEEE, 83*(3), 435–449.

Mamdani, E. H. (1977). Applications of fuzzy logic to approximate reasoning using linguistic synthesis. *IEEE Transaction on Computers, 26*(12), 1182–1191.

Kumar, K. S. (2009). Early software reliability and quality prediction (Ph.D. Thesis, IIT Kharagpur, Kharagpur, India).

Chapter 4
Multistage Model for Residual Fault Prediction

4.1 Introduction

Software reliability is defined as the probability of failure-free software operation for a specified period of time in a specified environment and is widely recognized as one of the most significant attributes of software quality (Lyu 1996). Over past decades, many software reliability growth models (SRGMs) have been presented to estimate important reliability measures such as the mean time to failure, the number of remaining faults, defect levels, and the failure intensity. Software reliability can be viewed form the two view points—user's view and developer's view. From a user's point of view, software reliability can be defined as the probability of a software system or component to perform its intended function under the specified operating conditions over the specified period of time. From developer's point of view, the reliability of the system can be measured as the number of residual faults that are likely to be found during testing or operational usage. This study aims to assure software reliability from developer's point of view.

The reliability of a software system depends on the number of residual faults sitting dormant inside. In fact, many of the software reliability models attempt to measure the number of residual bugs in the program (Goel and Okumoto 1979; Goel 1985). Problems with these models are that they require fault/failure data collected during testing. This is the main practical limitation of traditional models particularly when the software is being developed from scratch without any fault/failure data. One solution to this problem may be the use of software metrics for residual faults prediction. The terms "faults" and "residual faults" are synonymously used throughout this chapter.

Software metrics are used to describe the various activities concerned with software engineering measurements. One of the important activities is to produce a number that characterizes properties of the software for making various decisions. Further, metrics can be easily obtained using the expert of the domain for the various software development activities. These software metrics can be used to build management decision-supporting tools covering different aspects of software development and testing and empowering managers to carryout various useful

A. K. Pandey and N. K. Goyal, *Early Software Reliability Prediction*,
Studies in Fuzziness and Soft Computing 303, DOI: 10.1007/978-81-322-1176-1_4,
© Springer India 2013

prediction, assessments, and trade-offs during the software life cycle (Fenton and Neil 2000). Therefore, software metrics can be used to predict software reliability, quality, and resource requirements, in the absence of fault/failure data.

Software faults cause failures when executed and thus impact reliability and quality of software system adversely. Prediction of these software faults across the various phases of software development is desirable for both developers and project managers. Generally, these faults are introduced during software development and keep on propagating across the subsequent phases unless they are detected and corrected through testing or review process. Finally, the undetected and uncorrected faults get delivered with software, causing failures during software operation. In order to achieve high software reliability, the number of faults delivered in software should be kept at minimum level.

Fault density (some time also referred as defect density) is a measure of the total known faults divided by the size of the software entity being measured. More about fault density and its measurement have been discussed in Chap. 1. Fault density measure is the more desirable measure than number of faults for reliability and quality indicator (Pandey and Goyal 2010a). The term fault density indicator (FDI) is used in this approach to represent fault density at the end of each phase of software life cycle (Pandey and Goyal 2010a). FDI for requirements phase may be obtained by combining various requirements metrics (RM). Similarly, FDI for design, coding, and testing phases can be generated using design metrics (DM), coding metrics, and testing metrics, respectively. On the basis of FDI of testing phase, the number of residual faults is predicted.

This chapter proposes a multistage model for residual fault prediction without using software quality metrics and fuzzy inference system (FIS). The model utilizes software metrics and finds FDI for each development phase. On the basis of FDIs at the end of testing phase, the number of residual faults is predicted using a conversion equation. The proposed model assumes that the software is being developed through waterfall process model (Pressman 2005).

The chapter is organized as follows: Sect. 4.2 discusses about research backgrounds. Section 4.3 describes the proposed model. Section 4.4 discusses the implementation approach of the model using FIS. Section 4.5 provides a case study for residual fault prediction. Results are discussed in Sect. 4.6, whereas conclusions are presented in Sect. 4.7.

4.2 Research Background

4.2.1 Software Metrics

Quantitative measurement is necessary for better understanding and improving performance of engineering products. Hardware world has established its direct measurements such as mass, voltage, or temperature. Unfortunately, software world is still facing problem toward standardizations of measures and metrics. Due to this, most of the software measures and metrics are indirect.

IEEE Standard Glossary (IEEE 1991) defines metric as a "quantitative measures of the degree to which a system, components, or process possess a given attribute." Software metric is a measure of some attribute (internal or external) of software product, process, or resource. These software metrics can be broadly classified as follows: product metrics, process metrics, and resource metrics as shown in Table 4.1. Product metrics describe the characteristics of the product such as size, functionality, complexity, modularity, performance, reliability, maintainability, and quality. Process metrics can be used to improve software development and maintenance process and include time, effort, number of requirements change, cost effectiveness, etc. Resource metrics describe the resource characteristics and execution such as productivity, maturity, quality. There may be some overlap among these software metrics and some metrics may belong to multiple categories.

Software quality metrics are a subset of software metrics that focus on the quality aspects of the product, process, and resource (Kan 2002). In general, software quality metrics are more closely associated with process and product

Table 4.1 Classification of software metrics (Fenton and Neil 2000)

Entities	Attributes	
	Internal	External
Product		
Specifications	Size, reuse, modularity, functionality, correctness, etc	Comprehensibility, maintainability, etc
Design	Size, reuse, modularity, coupling, cohesiveness, etc	Quality, complexity, maintainability, etc
Code	Functionality, algorithmic complexity, control flow, etc	Reliability, usability, maintainability, etc
Test data	Size, coverage level, etc	Quality, reusability, etc
Processes		
Constructing specification	Time, effort, number of requirements changes, etc	Quality, cost, stability, etc
Detailed design	Time, effort, number of specification fault found, etc	Cost, cost effectiveness, etc
Testing	Time, effort, number of coding fault found, etc	Cost, cost effectiveness, stability, etc
Resources		
Personnel	Age, price, etc	Productivity, experience, intelligence, etc
Teams	Size, communication level, structuredness, etc	Productivity, quality, etc
Organizations	Size, ISO certification, CMM level, etc	Maturity, profitability, etc
Software	Price, size, etc	Usability, reliability, etc
Hardware	Price, speed, memory size, etc	Reliability, etc
Offices	Size, temperature, light, etc	Comfort, quality, etc

metrics than with resource metrics. Nonetheless, the resource parameters such as the team size, skill level of programmer, maturity level of organization, and working environments certainly affect the quality of the product. The quality must be viewed from the entire software life cycle perspective.

On reviewing various software metric datasets such as NASA (2004) and PROMISE (2007), it has been observed that the variation in software metric values follow either linear scale or logarithmic scale based on their nature. Therefore, software metrics are divided into two categories on the basis of their nature: (1) the variation in metric values that follows linear scale or (2) the variation in metric values that follows logarithmic. The metrics considered on linear scale shows considerable effects when there is some addition or reduction in the value, while metrics considered on logarithmic scale show significant variation when they are changed in multiplicity (ratios). On the basis of their nature, fuzzy profiles of software metrics are developed (Pandey and Goyal 2010a) and discussed in Sect. 4.4.2.

4.2.2 Fault Density Indicator

Fault density of a particular phase of software life cycle can be accurately estimated or predicted only if there is exact information about the number of faults as well as the size of that document. At the end of the phase, size can be estimated easily but the number of exact residual faults in the output document/code of the phase cannot be estimated. Thus, software metrics are used to estimate or predict number of fault, and so fault density.

A software engineer collects measure and develops metrics, so that indicators will be obtained. An indicator is a metric or combination of metrics that provides insight into software process, a software project, or the product itself (Pressman 2005). The term FDI is used in proposed approach to represent fault density at the end of each phase of software life cycle (Pandey and Goyal 2010a). FDI for requirements phase may be obtained by combining various RM. Similarly, FDI for design, coding, and testing phases can be generated using DM, coding metrics, and testing metrics, respectively.

4.3 Overview of the Proposed Model

Prediction of residual faults in the delivered software is highly desirable by the software industries. It provides an opportunity for the determination of software quality, reliability, cost overrun, and optimal development strategies. Software metrics can be measured across the development of software life cycle and can be used to predict the number of faults present in the software. Prior research has shown that software product and process metrics collected during the software development provide the basis for reliability predictions (Schneidewind 1992;

Khoshgoftaar and Munson 1990; Fenton and Neil 2000; Kumar and Misra 2008). A multistage model for residual fault prediction is proposed in this chapter using various software metrics and FIS.

The proposed model structure follows waterfall model, a well-known software development process model (Pressman 2005). Each stage of the model corresponds to the stage of waterfall model as shown in Fig. 4.1. The model structure is divided into five consecutive phases: I, II, III, IV, and V, representing requirement, design, coding, testing, and fault prediction phases, respectively. Requirements phase predicts the fault density indicator at the end of requirement phase (FDR) using relevant requirement metrics. Design phase predicts the fault density indicator at the end of design phase (FDD) using DM as well as output of the requirements phase (FDR). Similarly, coding phase uses coding metrics and design phase's output, that is, FDD to predict the fault density indicator at the end of coding phase (FDC). The testing phase predicts the fault density indicator of testing phase (FDT) using testing metrics as well as output of its previous phase, that is, FDC. Finally, on the basis of FDIs at the end of testing phase, the number of residual faults is predicted using a conversion equation.

Software metrics are taken from the PROMISE repository (2007). PROMISE repository is a software engineering repository dataset made publically available and shown in Appendix B. This repository contains twenty-seven software metrics, which are applicable to different phases of software development. Among these twenty-seven metrics, some metrics can also be applied to more than one

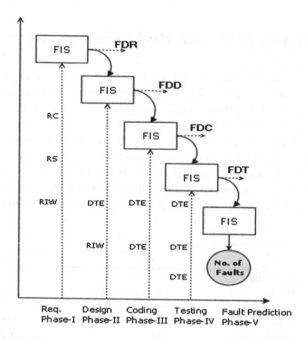

Fig. 4.1 Residual fault prediction model

development phase. For example, defined process followed (DPF) metric is applicable to all the development phases. Apart from that, some metrics such as quality of documentation inspected, programmer capability, and well defined testing process are redundant and do not provide any extra information. For example, relevant experience of specification and documentation staff, and relevant development staff experience are seem to convey almost same information. Therefore, only ten software metrics are selected which have more impact on software fault. Among these, three software metrics are identified to be more relevant at requirement phase, two are more relevant for design phase, again two are for coding phase, and three have been selected for coding phase.

Three RM: (a) requirements complexity (RC), requirements stability (RS), and review, inspection, and walkthrough (RIW) are selected at requirements phase. Similarly, at design phase, two DM are selected: design team experience (DTE) and process maturity (PM). Two coding phase metrics (CM), coding team experience (CTE) and DPF, are considered at coding phase. Three testing phase metrics (TM) selected are as follows: testing team experience (TTE), stakeholders involvement (SI), and size of the software (in KLOC) are taken as input at testing phase. The outputs of the model are fault density indicator at the end of requirements phase (FDR), design phase (FDD), coding phase (FDC), and testing phase (FDT). These metrics are given in Table 4.2.

In summary, the proposed model takes software metrics from PROMISE repository (PROMISE 2007), develops the fuzzy profiles of these metrics on the basis of their nature, and predicts FDIs, and residual faults using FIS.

4.3.1 Description of Software Metrics and their Nature

RC, RS, and RIW are found to be more suitable for requirement phase and influence requirements faults. If RC is more, the number of faults will be more, but this is not true for RS and the number of faults will reduce if the RS value is increased. Similarly, if there are more reviews and inspection, more faults will be detected and corrected leaving fewer faults in the requirements phase. For design phase, two metrics DTE and PM are considered because both of these metrics are responsible for error-free software design. For higher values of DTE and PM, there will be lower the number of design faults in the software. At coding phase, CTE and DPF metrics are found more suitable to affect the coding faults. In general, it is

Table 4.2 Phase-wise input/output variables

No.	Phase	Input variables	Output variables
1	Requirement	RC, RS, RIW	FDR
2	Design	FDR, DTE, PM	FDD
3	Coding	FDD, CTE, DPF	FDC
4	Testing	FDT, TTE, SI, SIZE	FDT
5	Fault prediction	FDT	Number of faults

found that if CTE and DPF are more, the number of faults will be less. Lastly, for testing phase, three metrics TTE, SI, and SIZE are taken which can influence the fault density at this phase.

RC: At the requirement phase, various requirements are collected and expressed into a common language understandable to both the project team and the stakeholders. The main problem with requirement gathering is that stakeholders are not very clear to their requirements and may have variety of requirements with conflicting goals. Therefore, a requirement complexity measure is used that will take the natural language requirements as input and associate a value of requirement complexity such as low, medium, or high. If the requirement complexity is more, there is a greater probability that the software will contain faults. RC value is measured qualitatively by conducting questionnaires with various domain experts, software professionals, and researchers. On reviewing various software metric datasets (NASA 2004; PROMISE 2007) and expert opinion, logarithmic variation in RC value is considered.

RS: Requirement stability is a metric used to organize, control, and track changes to the originally specified requirements. RS provides an indication of the completeness, stability, and understanding of the requirements. In other words, requirement stability can be obtained from the numbers and frequency of change request, and the amount of information needed to complete the requirements. A lack of RS is responsible for poor product quality, increased cost, and schedule slippage. Lower the requirement stability is, greater the number of faults is expected that remains in the software. Like RC, the RS variation is considered as logarithmic on the basis of reviews of various software metric datasets (NASA 2004; PROMISE 2007) and expert opinion.

RIW: Review, inspection, and walkthrough is the most powerful way to improve the quality of requirements documents. RIWs help us to discover defects and to ensure that product meets the stated requirements. During the review, work product is examined for defects by individuals other than the person who produced it. Reviews can be formal or informal. In informal reviews, roles are not defined and the process is ad hoc. The least formal reviews include hallway conversations, pair programming, and simply passing a product around. More formal reviews are planned and defined ahead of time, and their outcomes are used to improve development processes. Reviews can be conducted by individuals or groups. An inspection is the most formal type of group review. In a walkthrough, the producer describes the product and asks for comments from the participants. These gatherings generally serve to inform participants about the product rather than correct it. If the RIWs are more, there is a lower probability that the software will contain faults. RIW variation is considered to be linear in nature.

DTE: Design team experice is vital because of people factor. People gather requirements, people design software, and people write software for people. Software design creates a model upon which the software is build. Basically, this is most important software development activity where requirements are translated into a blueprint for constructing the software. The people involved in the software design must have a sound technical background and experience to produce a fault-free design. Therefore, DTE is considered as most important metric to indicate the quality of design. If DTE is more, the number of faults will be lesser in the design document. Also, DTE variation is also found to be effective on logarithmic scale.

PM: As discussed in Chap. 3, the capability maturity model (CMM) plays a major role in defining software process improvement. One of the widely adopted frameworks is the CMM developed by the software engineering institute (SEI) at Carnegie Mellon University (Paulk et al. 1993). Based on the specific software practices adopted, the CMM classifies the software PM into five maturity levels CMM level 1–5. There are many experts who argue that the "quality" of the development process is the best predictor of product quality and hence, by default, of residual faults (Fenton and Neil 1999). There are evidences that higher level of maturity deliver products with lower residual defect density. PM variation is considered of linear nature in this study.

CTE: Like design, the people involved in the software coding should have a sound technical background and experience to produce a high-quality program without any fault. Therefore, CTE is selected as a metric to indicate the quality of program. If CTE is more, the number of faults will be lesser in the design. CTE variation is also found to be effective on logarithmic scale.

DPF: DPF is very useful metric from management point of view to meet the target or desired software quality and reliability. This provides a means to understand the objectives of software project, its scope, and limitations by defining the process in writing well before the project started. Typically, a project process definition may include standard process adopted for software development, organizational process and policies, staff role, inspection and review method, etc. The goal of project process definition is to provide guidance and support to the project team member to produce quality product economically. The goal can be achieved only if defined process is followed by all team members. A defined process helps team members to understand their responsibilities better and learn quick from mistakes. A team that follows an ad hoc process is at the risk of producing faulty products and also violating organizational process requirements. DPF variation is considered of linear nature.

TTE: Software testing accounts for the largest percent of effort in software development. People involved in software testing are destructive in nature and try their best to find unexpected errors. Therefore, skills and experience of test team have a great impact on the test quality. A good test is one that has a high probability to uncover unexpected faults. Like DTE and CTE, TTE metrics also affect the number of residual faults. Lower the TTE is, higher the number of residual faults is expected. Like DTE and CTE, TTE also follows logarithmic in nature.

Stakeholders' involvements (SIs): Stakeholders are those persons, groups, individuals, and parties that are directly or indirectly affected by the software. No project will be complete without the support of its stakeholders. If we do not get the support from the key stakeholders, the project may be in risk. Not only support but also its degree is important to produce a high-quality software product. If the SIs are more, there is a high probability that the software will able to satisfy most of the stakeholders needs without failing. Thus, if SI is more, the number of faults will be lesser. SI variation is considered to be logarithmic in nature.

Size of the software (SIZE): Size is one of the most basic attributes of software. Traditional measure for software size is the number of line of code (LOC). Function points and feature points are can also be used to measure size of the software. Size is measured to normalize efforts, quality, and cost. In general, it has been found that larger and smaller software may contain more number of faults when compared to medium-sized software. SIZE variation is considered to be logarithmic in nature.

4.4 Model Implementation

The model is based on fuzzy logic and implemented in MATLAB. The model consists of the following steps:

1. Identification of independent and dependent variables.
2. Development of fuzzy profiles.
3. Developing fuzzy rules.
4. Information processing.
5. Residual fault prediction.

4.4.1 Independent and Dependent Variables

Identification of various independent and dependent variables plays an important role in developing a FIS. The independent variable is the variable that can be adjusted with the help of expert opinions and is hypothesized to influence the dependent variable. The dependent variable is the variable that is to be measured or predicted. It is the variable that reflects influence of the independent variables.

Software metrics are independent variables in our study. These are input variables to the model to derive dependent variables (output). Independent variables are taken from PROMISE repository as given in Table 4.3. Fault density indicators and the number of residual faults are the dependent variables in this study. There are four FDIs: FDR, FDD, FDC, and FDT corresponding to requirements, design, coding, and testing phases, respectively. Various dependent variables are given in Table 4.4. It is important to mention here that FDT works as independent variable for fault prediction.

4.4.2 Development of Fuzzy Profiles

Input/output variables identified at the previous stage are fuzzy in nature and characterized by fuzzy numbers. Fuzzy numbers are subset from the real numbers

Table 4.3 Independent variables

#	Independent variables
1	Requirements complexity (RC)
2	Requirements stability (RS)
3	Review, inspection, and walkthrough (RIW)
4	Design team experience (DTE)
5	Process maturity (PM)
6	Coding team experience (CTE)
7	Defined process followed (DPF)
8	Testing team experience (TTE)
9	Stakeholder involvement (SI)
10	Size of program in LOC (SIZE)

Table 4.4 Dependent variables

#	Dependent variables
1	Fault density indicator at requirements phase (FDR)
2	Fault density indicator at design phase (FDD)
3	Fault density indicator at coding phase (FDC)
4	Fault density indicator at testing phase (FDT)
5	Total number of residual faults predicted (Faults)

set, representing the uncertain values. All fuzzy numbers are related to degrees of membership, which state how true it is to say if something belongs or not to a determined set.

There are various types of fuzzy numbers, and its nomenclature is, in general, associated with its format, such as sine numbers, bell shape, polygonal, trapezoids, and triangular (Ross 2005). Triangular fuzzy numbers (TFN) are convenient to work with because in real space, adding two TFN involves adding the corresponding vertices used to define the TFNs. Similarly, simple formulas can be used to subtract or find the image of TFNs. Also, TFNs are well suited to modeling and design because their arithmetic operators and functions are developed, which allow fast operation on equations. Because of these properties, TFNs are considered for all input/output variables.

For the entire input variables, five linguistic levels are considered such as very low (VL), low (L), moderate (M), high (H), and very high (VH). Similarly, seven levels are assigned to all the output variables such as very very low (VVL), VL, L, medium (M), H, VH, and very very high (VVH).

The data, which may be useful for selecting appropriate linguistic variable, are generally available in one or more forms such as expert's opinion, user's expectations, and record of existing field data from previous release or similar systems. Fuzzy membership functions are generated utilizing the linguistic categories such as identified by a human expert to express his/her assessment. On the basis of their nature, fuzzy profiles of each software metrics are developed as follows:

a. For logarithmic scale nature software metrics,

$$\text{Fuzzy Profile} = \left[1 - \frac{\{\log_{10}(1 \text{ to } 5)\}}{\{\log_{10}(5)\}}\right]$$

The profiles will take the values as VL (0; 0; 0.14), L (0; 0.14; 0.32), M (0.14; 0.32; 0.57), H (0.32; 0.57; 1.00), and VH (0.57; 1.00; 1.00).

b. For linear scale nature software metrics,

$$\text{Fuzzy Profile} = \left[\frac{(0 \text{ to } 4)}{4}\right]$$

The profiles may take the values as VL (0; 0; 0.25), L (0; 0.25; 0.50), M (0.25; 0.50; 0.75), H (0.50; 0.75; 1.00), and VH (0.75; 1.00; 1.00).

c. Outputs are considered on logarithmic scale,

$$\text{Fuzzy Profile} = \left[1 - \frac{\{\log_{10}(1 \text{ to } 7)\}}{\{\log_{10}(7)\}}\right]$$

The profiles may take the values as VVL (0; 0; 0.08), VL (0; 0.08; 0.17), L (0.08; 0.17; 0.29), M (0.14; 0.32; 0.57), H (0.17; 0.29; 0.44), VH (0.44; 0.64; 1.00), and VVH (0.64; 1.00; 1.00).

As stated earlier, out of ten input variables, only three variables (RIW, PM, and DPF) are considered as linear scale and remaining seven variables follow logarithmic scale. All output variables are assumed to be following logarithmic scale. Fuzzy profile of various independent and dependent variable is shown in Figs. 4.2, 4.3, 4.4, 4.5, 4.6, 4.7, 4.8, 4.9, 4.10, 4.11, 4.12, 4.13, 4.14, 4.15.

4.4.3 Development of Fuzzy Rules

The most important part of the FIS is the rules, and how input variables interact with each other to generate results. In general, rules are formed using various

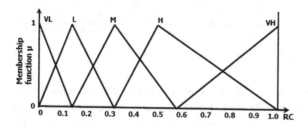

Fig. 4.2 Fuzzy profile of RC

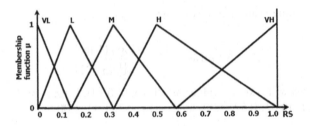

Fig. 4.3 Fuzzy profile of RS

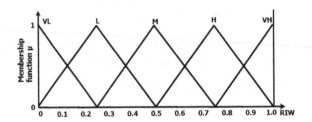

Fig. 4.4 Fuzzy profile of RIW

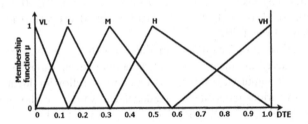

Fig. 4.5 Fuzzy profile of DTE

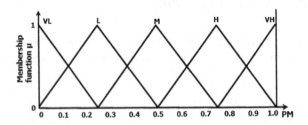

Fig. 4.6 Fuzzy profile of PM

domain experts, so that the system can emulate the inference of an actual expert. To develop fuzzy rule base, knowledge can be acquired from different sources such as domain experts, historical data analysis of similar or earlier system, and

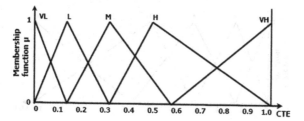

Fig. 4.7 Fuzzy profile of CTE

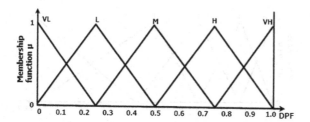

Fig. 4.8 Fuzzy profile of DPF

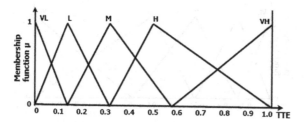

Fig. 4.9 Fuzzy profile of TTE

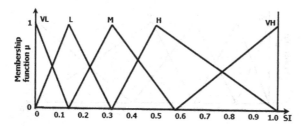

Fig. 4.10 Fuzzy profile of SI

engineering knowledge from existing literature's (Zhang and Pham 2000; Li and Smidts 2003). In presented model, some rules are generated from the software engineering point of view and some from project management view points. All the

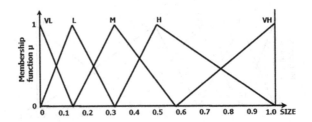

Fig. 4.11 Fuzzy profile of size

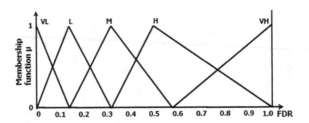

Fig. 4.12 Fuzzy profile FDR

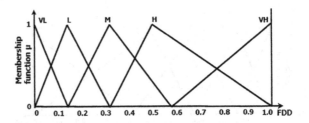

Fig. 4.13 Fuzzy profile of FDD

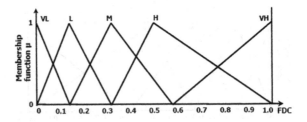

Fig. 4.14 Fuzzy profile of FDC

rules take the form of "If A then B." Tables 4.5, 4.6, 4.7, 4.8 show the fuzzy if–then rules required for each phase of software life cycle.

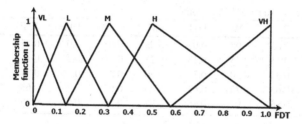

Fig. 4.15 Fuzzy profile of FDT

Table 4.5 Fuzzy rules at requirements phase

Rule	RC	RS	RIW	FDR
1	L	L	L	VL
2	L	L	M	L
3	L	L	H	M

Table 4.6 Fuzzy rules at design phase

Rule	FRP	DTE	PM	FDD
1	VL	L	L	VL
2	VL	L	M	VL
3	VL	L	H	L

Table 4.7 Fuzzy rules at coding phase

Rule	FDP	CTE	DPF	FDC
1	VL	L	L	VL
2	VL	L	M	VL
3	VL	L	H	L

Table 4.8 Fuzzy rules at testing phase

Rule	FDC	TTE	SI	SIZE	FDT
1	VL	VL	L	L	VL
2	VL	VL	L	M	VL
3	VL	VL	L	H	L

4.4.4 Information Processing

At this step, the fuzzy system maps all inputs on to an output. This process of mapping is known as fuzzy inference process or fuzzy reasoning. The basis for this mapping is the number of fuzzy if–then rules, and each of which describes the local behavior of the mapping. The Mamdani FIS (Mamdani 1977) is considered

Table 4.9 Centroid value of fuzzy profiles

	RC	RS	RIW	DTE	PM	CTE	DPF	TTE	SI	Size
VL	0.05	0.05	0.08	0.05	0.08	0.05	0.08	0.05	0.05	0.05
L	0.15	0.15	0.25	0.15	0.25	0.15	0.25	0.15	0.15	0.15
M	0.34	0.34	0.50	0.34	0.50	0.34	0.50	0.34	0.34	0.34
H	0.63	0.63	0.75	0.63	0.75	0.63	0.75	0.63	0.63	0.63
VH	0.86	0.86	0.92	0.86	0.92	0.86	0.92	0.86	0.86	0.86

here for all the information processing. Centroid value of each fuzzy profile is taken for computation purpose. Centroid value of each fuzzy profile can be computed from its fuzzy number, and these are shown in the Table 4.9.

Another vital part of information processing is the defuzzification process, which derives a crisp value from a number of fuzzy values. Various defuzzification techniques (Ross 2005) are available such as centroid, bisector, middle of maximum, largest of maximum, and smallest of maximum. The most commonly used method is the centroid method, which returns the center of area under the curve. In this study, centroid defuzzification method is considered.

4.4.5 Residual Fault Prediction

On the basis of FDIs at the end of testing phase, the total number of faults in the software is predicted. The faults are predicted from FTP value, which is an indicator of fault density at the end of testing phase. The number of faults detected is proportional to the size of software and the FDI value; therefore, residual faults (FaultP) can be found as follows:

$$\text{FaultP} = C_1 \times \text{LOC} \times \text{FTP}$$

However, fault detection process is not exactly linear with size. As size of a software increases, portion of faults detected decreases due to saturation, time, and efforts requirements, increased possibility due to interactions between variable and instructions. Therefore, the FTP value is divided by $(1 + \exp(-\text{LOC}i/C_2))$. The C_2 value scales the effect of LOC value at which saturation is expected. Thus, residual fault for project i is predicted as follows:

$$\text{FaultP}i = C_1 \times \text{LOC}i \times \text{FTP}/(1 + \exp(-\text{LOC}i/C_2))$$

where FaultPi is the total number of predicted residual faults in the ith software project, LOCi is the size of ith software project, FTPi is the FDI at the end of testing phase of project i, C1 and C2 are constants obtained through learning.

Value of C_1 and C_2 are obtained from available project data. The value of C_2 is taken as 10^7 considering that this value is large enough for any software to reach saturation. Value of C_1 is found, using LOC, Faults, and FTP data for odd numbered project from the list, such that mean square error (MSE) is minimized. The value of C_1 is obtained as 0.04.

The proposed fuzzy inference model uses generic fuzzy profiles to capture variation in faults present in the software through FDIs, while the remaining faults are obtained through conversion equation considering learning from existing project data.

4.5   Case Study

4.5.1   Dataset Used

In order to validate the proposed model, *qqdefects* dataset is used. This dataset is obtained from PROMISE Software Engineering Repository dataset which is made publicly available in order to encourage repeatable, verifiable, refutable, and/or improvable predictive models of software engineering (PROMISE 2007). PROMISE dataset (*qqdefects*) consists of qualitative and quantitative data about 31 projects from a consumer electronics company. These project data are given in Appendix B. These qualitative attributes and quantitative data are measured on a five-point ranking scale such as VL, L, M, H, and VH.

4.5.2   Metrics Considered in "qqdefects" Dataset

Each project of *qqdefects* dataset consists of thirty software metrics corresponding to different phases of software life cycle (see Appendix B). Only ten metrics [F1 (RC), S7 (RS), S3 (RIW), D1 (DTE), P9 (PM), D2 (CTE), D3 (DPF), T2 (TTE), P5 (SI), and K (SIZE)] are considered for the present study. These metrics are listed in Table 4.3. These metrics are used as independent variables (also called input variable) and used for deriving dependent variables (also called output variables) using FIS and the methods discussed in Sect. 4.4. The rest of metrics are not considered to make large impact on fault prediction as their information is either already contained in the selected metrics or their effect on number of faults is very less compared to the metrics selected.

4.5.3   Conversion of Dataset

Datasets of *qqdefects* are given on a five-point ranking scale such as VL, L, M, H, and VH. To use these data in FIS, data are fuzzified using method discussed in Sect. 4.4.2. After fuzzification, centroid value is taken of each fuzzy profile for further computation of FDI and residual faults. Centroid value of metrics for each fuzzy profile is shown in the Table 4.11.

Table 4.10 Twenty-two project dataset of "*qqdefects*"

#	Project	RC F1	RS S7	RIW S3	DTE D1	PM P9	CTE D2	DPF D3	TTE T2	SI P5	SIZE K	Faults TD
1	1	M	L	VH	L	H	H	H	H	H	6.02	148
2	2	L	H	VH	L	H	H	H	H	H	0.90	31
3	3	H	H	VH	H	VH	VH	H	H	VH	53.86	209
4	5	H	M	H	L	H	M	H	M	M	14.00	373
5	7	L	M	VH	M	H	VH	H	M	VH	21.00	204
6	8	M	H	H	H	M	H	M	M	H	5.79	53
7	10	M	H	H	H	H	H	H	M	H	4.84	29
8	11	H	H	H	H	H	H	H	H	H	4.37	71
9	12	H	L	H	VH	H	M	M	H	H	19.00	90
10	13	H	L	M	H	H	H	H	M	H	49.10	129
11	14	VH	H	H	H	H	H	H	H	H	58.30	672
12	15	H	VL	H	H	H	H	H	H	VH	154.00	1,768
13	16	L	M	H	H	H	H	H	H	VH	26.67	109
14	17	L	M	M	M	H	M	H	L	M	33.00	688
15	19	H	M	H	H	H	H	H	M	H	87.00	476
16	20	VH	VL	M	VL	H	VL	L	VL	H	50.00	928
17	21	L	M	H	H	H	H	H	H	H	22.00	196
18	22	M	L	M	H	H	M	L	M	H	44.00	184
19	23	H	M	VH	L	H	H	H	H	H	61.00	680
20	27	H	M	VH	M	H	L	M	M	M	52.00	412
21	29	M	VH	VH	VH	H	VH	H	VH	VH	11.00	91
22	30	L	VH	VH	H	H	H	H	H	VH	1.00	5

There are total 31 project data in the PROMISE database as shown in sheet (Appendix B). Five projects data (4, 6, 9, 25, and 26) have incomplete data, that is, some attribute values are missing. Furthermore, four projects (18, 24, 28, and 31) have shown unexpected behaviors and therefore not considered for the present study. Finally, the project dataset is reduced to twenty-two as shown Table 4.10. These twenty-two project data are further converted using Table 4.9 to represent the dataset with their centroid value as shown in Table 4.11.

4.6 Results and Discussion

After converting the project data into the appropriate form, we provide these data to the FISs to get output as shown in Table 4.13. FDR, FDD, FDC, and FDT are the fault densities (FDIs) of requirements, design, coding, and testing phases, respectively. The number of residual faults is predicted using FDT value as discussed in Sect. 4.4.5. Prediction results are sorted on the basis of project size in order to visualize the effect of project size on prediction accuracy.

Table 4.11 Project dataset with centroid value

#	Project	RC	RS	RIW	DTE	PM	CTE	DPF	TTE	SI	SIZE	Faults
1	1	0.34	0.15	0.92	0.15	0.75	0.63	0.75	0.63	0.63	0.15	148
2	2	0.15	0.63	0.92	0.15	0.75	0.63	0.75	0.63	0.63	0.15	31
3	3	0.63	0.63	0.92	0.63	0.92	0.86	0.75	0.63	0.86	0.86	209
4	5	0.63	0.34	0.75	0.15	0.75	0.34	0.75	0.34	0.34	0.34	373
5	7	0.15	0.34	0.92	0.34	0.75	0.86	0.75	0.34	0.86	0.63	204
6	8	0.34	0.63	0.75	0.63	0.5	0.63	0.5	0.34	0.63	0.15	53
7	10	0.34	0.63	0.75	0.63	0.75	0.63	0.75	0.34	0.63	0.15	29
8	11	0.63	0.63	0.75	0.63	0.75	0.63	0.5	0.63	0.63	0.15	71
9	12	0.63	0.15	0.75	0.86	0.75	0.34	0.5	0.63	0.63	0.34	90
10	13	0.63	0.15	0.5	0.63	0.75	0.63	0.75	0.34	0.63	0.63	129
11	14	0.86	0.63	0.75	0.63	0.75	0.63	0.75	0.63	0.63	0.86	672
12	15	0.63	0.05	0.75	0.63	0.75	0.63	0.75	0.63	0.86	0.86	1,768
13	16	0.05	0.34	0.75	0.63	0.75	0.63	0.75	0.63	0.86	0.63	109
14	17	0.05	0.34	0.5	0.34	0.75	0.34	0.75	0.15	0.34	0.63	688
15	19	0.63	0.34	0.75	0.63	0.75	0.63	0.75	0.34	0.63	0.86	476
16	20	0.86	0.05	0.5	0.05	0.75	0.05	0.25	0.05	0.63	0.86	928
17	21	0.05	0.34	0.75	0.63	0.75	0.63	0.75	0.63	0.63	0.63	196
18	22	0.34	0.15	0.5	0.63	0.75	0.34	0.25	0.34	0.63	0.63	184
19	23	0.63	0.34	0.92	0.15	0.75	0.63	0.75	0.63	0.63	0.86	680
20	27	0.34	0.34	0.92	0.34	0.75	0.15	0.5	0.34	0.34	0.86	412
21	29	0.34	0.86	0.92	0.86	0.75	0.86	0.75	0.86	0.86	0.34	91
22	30	0.05	0.86	0.92	0.63	0.75	0.63	0.75	0.63	0.86	0.15	5

The FDR, FDD, FDC, FDT, the number of residual faults predicted by the proposed model, and faults predicted by Fenton et al. (2008) are provided in Table 4.12. The prediction results are sorted on the basis of project size just to visualize the effect of size the residual faults.

In order to validate the prediction accuracy of the proposed model, we have used the following evaluation measures which are quite common to validate such type of prediction models.

1. Error $(E) = [F_A - F_P]$, where F_A and F_P are the number of actual and predicted residual faults, respectively.

2. Cumulative Error (CE) $= \sum_{i=1}^{n} E$

3. Mean Error (ME) $= \frac{1}{n} \times \sum_{i=1}^{n} E$

4. Sum of Square Error (SSE) $= \sum_{i=1}^{n} (F_A - F_P)^2$

5. Mean Square Error (MSE) $= \frac{1}{n} \times \sum_{i=1}^{n} (F_A - F_P)^2$

6. Root Mean Square Error (RMSE) $= \sqrt{MSE}$

Table 4.12 Fault densities, actual faults, and faults predicted

#	FDR	FDD	FDC	FDT	Size (LOC)	Actual faults	Proposed model	Fenton et al. (2008)
1	0.09	0.19	0.14	0.30	900	31.00	5.48	52.00
2	0.08	0.08	0.07	0.30	1,000	5.00	6.02	46.00
3	0.46	0.30	0.18	0.46	4,370	71.00	40.61	51.00
4	0.18	0.14	0.13	0.27	4,840	29.00	26.17	203.00
5	0.18	0.30	0.18	0.46	5,790	53.00	53.81	48.00
6	0.32	0.46	0.39	0.46	6,020	148.00	55.94	75.00
7	0.21	0.19	0.14	0.50	11,000	91.00	110.06	116.00
8	0.63	0.66	0.68	0.83	14,000	373.00	232.48	349.00
9	0.29	0.25	0.19	0.46	19,000	90.00	176.25	347.00
10	0.19	0.28	0.17	0.27	21,000	204.00	113.43	262.00
11	0.19	0.14	0.13	0.24	22,000	196.00	105.51	259.00
12	0.19	0.14	0.13	0.24	26,670	109.00	127.94	145.00
13	0.19	0.28	0.17	0.21	33,000	688.00	135.74	444.00
14	0.57	0.30	0.30	0.33	44,000	184.00	291.02	501.00
15	0.46	0.30	0.23	0.34	49,100	129.00	336.59	516.00
16	0.72	0.73	0.84	0.87	50,000	928.00	869.23	986.00
17	0.42	0.43	0.18	0.39	52,000	412.00	400.17	430.00
18	0.29	0.26	0.16	0.20	53,860	209.00	210.61	210.00
19	0.46	0.30	0.23	0.60	58,300	672.00	697.48	674.00
20	0.48	0.49	0.39	0.56	61,000	680.00	690.17	722.00
21	0.63	0.31	0.25	0.33	87,000	476.00	573.44	581.00
22	0.46	0.30	0.23	0.53	154,000	1,768.00	1,650.89	1,526.00

7. Mean Percent Error (MPE) $= \frac{1}{n} \times \sum_{i=1}^{n} \left(\frac{F_A - F_P}{F_A} \right) \times 100$

8. Mean Absolute Percent Error (MAPE) $= \frac{1}{n} \times \sum_{i=1}^{n} \left| \frac{F_A - F_P}{F_A} \right| \times 100$

Proposed model prediction result is compared with a model using Bayesian nets (Fenton et al. 2008). Table 4.13 summarizes the comparative value of various evaluation measures. From the table, it is clear that proposed approach, which is based on FIS, provides more accurate results than the model based on Bayesian nets provided by Fenton et al. (2008). In their study, Fenton et al. (2008) have used

Table 4.13 Comparative values of the model results

Evaluation measures	Proposed approach	Fenton et al. (2008) approach
CE	641.00	−1,041.00
ME	29.14	−47.32
SSE	440,943.00	500,895.00
MSE	20,042.86	22,767.95
RMSE	141.57	150.89
MPE	1.09	−103.86
MAPE	37.49	116.81

all the thirty-one *qqdefects* project data (PROMISE 2007). This study have used only twenty-two project data, because five projects data have missing value with the corresponding metrics and four project data have been dropped as they are showing unexpected behavior during the experiments.

Table 4.13 shows evaluation measures such as cumulative errors (CE), mean error (ME), sum of square error (SSE), MSE, root mean square error (RMSE), mean percent error (MPE), and mean absolute percent error (MAPE). CE value of the proposed model is found to be 641.00 when compared to the −1041.00, which is the CE value given by of the Fenton et al. (2008). Similarly, proposed model results ME value of 29.14 when compared to −47.32 of Fenton et al. (2008). SSE value of the proposed approach is also found to be 441454 which are better than Fenton et al. (2008) value 500895. The MSE and RMSE values for the proposed model are found to be 20042.86 and 141.57, respectively. The MSE and RMSE values for the Fenton et al. (2008) model are found to be 22767.95 and 150.89, respectively. Finally, MPE and MAPE values are also calculated for the proposed model and found to 1.09 and 37.49, respectively. These MPE and MAPE values are found better than Fenton et al. (2008) values as −103.86 and 116.81.

4.7 Summary

This chapter has presented a multistage model that predicts FDI for each development phase. Using the FDT, existing residual faults in the software can be predicted with reasonable accuracy. The model has considered various software metrics given in the PROMISE repository (2007), developed the fuzzy profiles of these metrics on the basis of their nature, and predicted the number of residual faults using FIS. This model has considered all the development phases and predicted FDIs at the end of each phase of software development using relevant software metrics and FIS which is generic in nature. On the basis of FDIs at the end of testing phase, the number of residual faults is predicted using a conversion equation.

The prediction accuracy is compared to another model for the same dataset and found to be better. For software professionals, this model will provide an insight toward software metrics and its impact on software fault density during the development process.

References

Fenton, N. E., & Neil, M. (1999). A critique of software defect prediction models. *IEEE Transaction on Software Engineering, 25*(5), 675–689.

Fenton, N. E., & Neil, M. (2000), Software metrics: Roadmap, *Proceedings of the Conference on the Future of Software Engineering*, (pp. 375–370). Limerick, Ireland.

Fenton, N., Neil, N., Marsh, W., Hearty, P., Radlinski, L., & Krause, P. (2008). On the effectiveness of early life cycle defect prediction with Bayesian nets. *Empirical of Software Engineering, 13*, 499–537.

Goel, A. L. (1985). Software reliability models: Assumptions, limitations, and applicability. *IEEE Transaction on Software Engineering, SE–11*(12), 1411–1423.

Goel, A. L., & Okumoto, K. (1979). A Time-dependent error detection rate model for software reliability and other performance measure. *IEEE Transaction on Reliability, R-28*, 206–211.

IEEE (1991). IEEE standard glossary of software engineering terminology, STD-729-991, ANSI/ IEEE.

Kan, S. H. (2002). *Metrics and models in software quality engineering* (2nd ed.). Reading, MA: Addison Wesley.

Khoshgoftaar, T. M., & Munson, J. C. (1990). Predicting software development errors using complexity metrics. *IEEE Journal on Selected Areas in Communication, 8*(2), 253–261.

Kumar, K. S., & Misra, R. B. (2008). An enhanced model for early software reliability prediction using software engineering metrics, *Proceedings of 2nd International Conference on Secure System Integration and Reliability Improvement*, (pp. 177–178).

Li, M., & Smidts, C. (2003). A ranking of software engineering measures based on expert opinion. *IEEE Transaction on Software Engineering, 29*(9), 811–824.

Lyu, M. R. (1996). *Handbook of Software Reliability Engineering*. NY: McGraw–Hill/IEE Computer Society Press.

Mamdani, E. H. (1977). Applications of fuzzy logic to approximate reasoning using linguistic synthesis. *IEEE Transaction on Computers, 26*(12), 1182–1191.

NASA (2004), NASA metrics data program, http://mdp.ivv.nasa.gov/.

Pandey, A. K., & Goyal, N. K. (2010). Multistage fault prediction model using process level software metrics. *International Journal of Communications in Dependability and Quality Management, 13*(1), 54–66.

Paulk, M. C., Weber, C. V., Curtis, B., & Chrissis, M. B. (1993). Capability maturity model version 1.1. *IEEE Software, 10*(3), 18–27.

Pressman, R. S. (2005). *Software engineering: A practitioner's approach* (6th ed.). New York: McGraw-Hill Publication.

PROMISE repository (2007). http://promisedata.org/repository.

Ross, T. J. (2005). *Fuzzy logic with engineering applications* (2nd ed.). India: Wiley.

Schneidewind, N. F. (1992). Methodology for validating software metrics. *IEEE Transactions on Software Engineering, 18*(5), 410–422.

Zhang, X., & Pham, H. (2000). An analysis of factors affecting software reliability. *The Journal of Systems and Software, 50*(1), 43–56.

Chapter 5
Prediction and Ranking of Fault-Prone Software Modules

5.1 Introduction

Large and complex software systems are developed by integrating various independent modules. It is important to ensure quality of these modules through independent testing where modules are tested and faults are removed as soon as failures are experienced. System failures due to the software failure are common and result in undesirable consequences. Moreover, it is difficult to produce fault-free software due to problem complexity, complexity of human behavior, and resource constraints.

A software system consists of various modules and, in general, it is known that a small number of software modules are responsible for majority of the failures (Ohlsson and Alberg 1996). Pareto principle can also be applied to the identification of software quality problems. Pareto principle says that most of the quality problems are due to a small percentage of the possible causes. Pareto analysis is commonly referred to as the "80–20" principle, that is, 20% of the causes account for 80% of the failures. If these 20% faulty modules are identified before testing, the software quality can be improved more cost-effectively, by allocating testing resource optimally. Therefore, it is desirable to classify the software module as fault-prone (FP) or not fault-prone (NFP) before testing. Once the modules are classified as FP or NFP, more testing efforts can be put on the FP module to produce reliable software. Furthermore, by ranking FP modules on the basis of its degree of fault-proneness; testing efforts can be optimized to achieve cost-effectiveness.

The quality of modules is judged on the basis of number of faults lying dormant in the module. A fault is a defect in a software module that causes failure when executed (Musa et al. 1987). A software module is said to be FP, when there is a high probability of finding faults during its operation. In other words, a FP software module is the one containing more number of expected faults than a given threshold value. The threshold value can take any positive value and depends on the project-specific criteria. In general, when software modules have been identified as FP, testing efforts are applied accordingly to improve their quality and reliability.

A. K. Pandey and N. K. Goyal, *Early Software Reliability Prediction*,
Studies in Fuzziness and Soft Computing 303, DOI: 10.1007/978-81-322-1176-1_5,
© Springer India 2013

It is important to note that all the modules are neither equally important nor do they contain an equal amount of faults. Also, testing time and resources are limited. Therefore, the testing resources should be allocated to modules based on their degree of fault-proneness. In other words, it is undesirable to apply equal amount of testing resource to all the software modules. Software modules may be either FP or NFP. Obviously, FP modules require more testing resources than NFP modules to improve its reliability. As stated earlier, FP modules are those which contains more number of expected faults than a given threshold value. There may be a possibility that two FP modules do not contain equal amount of faults. It is desirable to further rank these FP modules on the basis of their FP degree. A FP module can be ranked on the basis of its degree of fault-proneness using software metrics and fuzzy ordering algorithm (Ross 2005). After then, the testing resources should be allocated accordingly.

The variety of software quality classification models has been proposed by several authors (Fenton and Neil 1999; Khoshgoftaar and Allen 1999, 2000, 2003; Khoshgoftaar and Seliya 2003). The main problem with the traditional module prediction model is that they are all using the crisp values of software metrics and classify the module as a FP or NFP. It has been found that early-phase software metrics have fuzziness in nature and crisp value assignment seems to be impractical. Also, all the traditional models are predicting the module as either FP or NFP (i.e., we can use a crisp value "1" for representing FP and value "0" representing NFP). This type of prediction may suffer from some ambiguity and not desirable where testing resources are to be allocated on the basis of its degree of fault-proneness. It is unfair to allocate equal amount of testing resource to all FP or NFP modules. A software module cannot be 100% FP or NFP. Some degree of fault-proneness is associated with each module. This enables the tester to allocate testing resources accordingly to quality software in cost-effective manner.

In this chapter, a model for prediction and ranking of FP software module using data mining techniques (Han and Kamber 2001) and fuzzy inference modeling is proposed. The software modules are classified as FP or NFP through fuzzy inference system developed using a well-known classification algorithm ID3 (Iterative Dichotomizer 3). Further, using software metrics and fuzzy ordering algorithm (Ross 2005), FP modules are ranked on the basis of their degree of fault-proneness. The contribution of the proposed model includes the following:

- The best design, coding, and testing approach can be selected from various available alternatives after identifying the FP modules.
- Proposed model fulfills the two main objectives of software engineering (1) increase the quality of the software by identifying the FP and NFP modules (2) decrease the overall testing costs by prioritizing FP modules.
- Proposed model will help reliability and quality engineer to focus their limited testing resources on parts of the system likely to contain more faults.

Rest of the chapter is organized as follows: next section presents research backgrounds. Section 5.3 discusses the proposed model. Section 5.4 presents the model implementation. Section 5.5 presents proposed algorithms. Section 5.6 presents a case study. Section 5.7 contains results and discussion whereas summary is presented in Sect. 5.8.

5.2 Research Background

5.2.1 Data Mining

Today's advanced information systems have enabled collection of increasingly large amounts of data. To analyze these data, the interdisciplinary field of knowledge discovery in databases (KDD) has emerged (Han and Kamber 2001) as a useful technique. KDD comprises of many steps, namely data selection, data preprocessing, data transformation, data mining, and data interpretation and evaluation. Data mining forms a core activity in KDD.

Data mining entails the overall process of extracting knowledge from large amounts of data. Different types of data mining techniques are discussed in the literature such as regression, classification, and associations. Regression and classification are predictive data mining tasks whereas association rule mining is a descriptive data mining. The focus here is on classification technique, which is the task of classifying the data into predefined classes to its predictive characteristics. The result of a classification technique is a model which makes it possible to classify future data points based on a set of specific characteristics in an automated way. In the literature, there are many classification techniques, some of the most commonly used being ID3, C4.5, logistic regression, linear and quadratic discriminant analysis, k-nearest neighbor, artificial neural networks (ANN), and support vector machines (SVM) (Han and Kamber 2001). These techniques have been successfully applied in different domains like breast cancer detection in the biomedical sector, market basket analysis in the retail sector, and credit scoring in the financial sector. This paper, however, focuses on the use of data mining to support and improve the quality and reliability of software.

The main goal of software engineering is to produce high-quality software with low cost. To achieve these goals, software managers are continuously keeping track of faults within a software module as well as the budgetary and schedule constraints. Software data mining aims to tackle some of these issues by extracting knowledge from past project data that may be applied to future projects or development stages. It is a rapidly evolving domain of data mining and improves reliability and quality of software development by mining software repositories with comprehensible data mining techniques. In other words, software data mining mainly focuses on the prediction of faults in software modules by the use of data mining techniques. This enables software project managers to achieve their goals of developing reliable and quality software.

5.2.2 Software Metrics

Quality of a software module are greatly affected by the factors such as its size, complexity, development process involved, experience of the developer, etc. Software metrics can be categorized as product, process, and resource metrics (Fenton 1991). Software size metrics may also be classified on the basis of the software size such as total number line of code (LOC), executable LOC, and number of comment LOC. The value of these software metrics can be calculated statically, without program execution. For this, Halstead's software science had provided an analytical technique to measure size, development effort, and the cost of software product. For this, he computed four primitive program parameter and developed expressions to calculate program length, development effort, and time. The four primitive parameters are number of unique operators used in the program (n1), number of unique operands used in the program (n2), total number of operators used in the program (N1), and total number of operands used in the program (N2).

Three complexity metrics such as cyclomatic complexity (CC), essential complexity (EC), and design complexity (DC) are developed by MaCabe to indicate the complexity of software. CC is computed using the control flow graph of the program. The EC of a program calculated by first removing structured programming primitives from the program's control flow graph until the graph cannot be reduced any further, and then calculating the CC of the reduced graph. DC of a flow graph is the CC of its reduced graph. Reduction is performed to eliminate any complexity which does not influence the interrelationship between design modules. A number of studies have reported a positive correlation between program complexity and defects (Khoshgoftaar and Munson 1990). Modules with high complexity tend to contain the more number of defects in the module.

Apart from complexity metric, there is a branch count (BC) metric which tells the number of branches for each module. Branches are defined as those edges that exit from a decision node. The greater the number of branches in a program's modules, greater will be the testing resources required. In this study, KC2 dataset, which is a public domain defect dataset from the NASA metrics data program (MDP), and PROMISE repository (NASA 2004) are utilized. This dataset contains static code metrics (e.g., Halstead, McCabe, and LOC measures) along with number of defects.

5.2.3 Fuzzy Set Theory

This section presents a brief overview of concepts related to fuzzy set theory that are relevant to the present study (Zadeh 1965; Ross 2005).

Crisp and Fuzzy Sets: Crisp or classical set can be defined as a collection of well-defined distinct object. In other words, crisp sets contain objects that satisfy precise properties of membership. For crisp set, an element x in the universe X is either a member of some crisp set (A) or not. This binary issue of membership can be represented by a characteristic function as

$$\chi_A(x) = \begin{cases} 1, & \text{if } x \in A \\ 0, & \text{if } x \notin A \end{cases}$$

$\chi_A(x)$, provides an unambiguous membership of the element, x in a set A. A fuzzy set is a set containing elements that have varying degree of membership in the set. Unlike crisp set, elements in a fuzzy set need not be complete and can also be member of other fuzzy sets on the same universe. In this way, fuzzy set allows partial membership as well as binary membership.

Let $\tilde{A}$ is a fuzzy set of A, if an element in the universe, say, x is a member of fuzzy set $\tilde{A}$; then, this mapping is given by a membership function $\mu_{\tilde{A}}(x)$. The membership function $\mu_{\tilde{A}}(x)$ gives the degree of membership for each element in the fuzzy set $\tilde{A}$ and is defined in range [0, 1], where 1 represents elements that are completely in $\tilde{A}$, 0 represents elements that are not in $\tilde{A}$, and values between 0 and 1 represent partial membership in $\tilde{A}$. Formally, a fuzzy set $\tilde{A}$ can be represented using Zadeh's notation 4 as

$$\tilde{A} = \left\{ \frac{\mu_1}{x_1} + \frac{\mu_2}{x_2} + \frac{\mu_3}{x_3} + \cdots + \frac{\mu_n}{x_n} \right\}$$

where μ_1, μ_2, ..., μ_n are the membership values of the elements x_1, x_2, ..., x_n, respectively, in the fuzzy set $\tilde{A}$.

Crisp and fuzzy ordering: Decisions are sometimes made on the basis of rank, or ordinal ranking: which issue is best, which is second best, and so forth. For issues or actions that are deterministic, such as $X_1 = 10$, $X_2 = 20$, and $X_1 \geq X_2$, there is no ambiguity in the ranking and can be referred as crisp ordering. In situations where the issues or action are associated with some uncertainty, rank ordering may be ambiguous. This ambiguity or uncertainty can be handled by fuzzy ordering (Ross 2005). Let $\tilde{I}_1$ and $\tilde{I}_2$ are two fuzzy sets. Fuzzy set $\tilde{I}_1$ is greater than $\tilde{I}_2$ if the following condition satisfies,

$$T(\tilde{I}_1 \geq \tilde{I}_2) = \max_{(x_1 \geq x_2)} \left\{ \min\left(\mu_{\tilde{I}_1}(x_1), \mu_{\tilde{I}_2}(x_2) \right) \right\}$$

where, $T(\tilde{I}_1 \geq \tilde{I}_2)$ is the truth value on the interval [0, 1] and $\mu_{\tilde{I}_1}(x_1)$, $\mu_{\tilde{I}_2}(x_1)$ represents the degree of membership of first element in the fuzzy set $\tilde{I}_1$ and $\tilde{I}_2$, respectively. The definition expressed above is applicable for two fuzzy numbers and can be extended to the more general case for many fuzzy sets. Suppose, there are k fuzzy sets as $\tilde{I}_2$, $\tilde{I}_2$, ..., $\tilde{I}_k$. Then, the truth values of a specified ordinal ranking can be given as

$$T(I \geq \tilde{I}_1, \tilde{I}_2, \ldots \tilde{I}_k) = \left[T(\tilde{I} \geq \tilde{I}_1) \wedge T(\tilde{I} \geq \tilde{I}_2) \wedge \ldots \wedge T(\tilde{I} \geq \tilde{I}_k) \right]$$

$$T(I \geq \tilde{I}_1, \tilde{I}_2, \ldots \tilde{I}_k) = \min\left[T(\tilde{I} \geq \tilde{I}_1), T(\tilde{I} \geq \tilde{I}_2), \ldots \wedge (\tilde{I} \geq \tilde{I}_k) \right]$$

5.3 Proposed Model

Prediction of software faults is desirable for both software engineer as well as project manager. It provides an opportunity for the early identification of software quality, reliability, cost overrun, and optimal development strategies. Software metrics can be measured across the development of software life cycle and can be used to predict the number of faults present in the software. Prior research shows that software product and process metrics collected during the software development provide basis for reliability predictions (Schneidewind 1992; Khoshgoftaar and Munson 1990; Fenton and Neil 2000; Kumar and Misra 2008).

5.3.1 Assumptions and Architecture of the Model

A1:   It is assumed that knowledge is stored in software metrics which helps in quality prediction at early stage as the quality information of the developing software is not available

A2:   Previous software project data of the similar domain will provide a good training to the model

A3:   It is also assumed that decision tree induction algorithms (ID3) are efficient classification algorithms for the purpose of fault prediction

A3:   Fuzzy profile of each software metrics of the modules can be obtained

The proposed model structure is divided into three parts, viz. learning, classification and ranking as shown in the Fig. 5.1. Learning is meant for providing training to the model using some training dataset. In this study, a part of KC2 dataset is used for training that will be discussed in Sect. 5.6.1. Classification provides a way to classify the dataset into two different classes using classification algorithm. ID3 algorithm is used for generating a classification tree, which classifies KC2 dataset into two different classes as FP module or NFP. Classification procedure and decision tree construction are discussed in Sect. 5.4.2. Ranking is concerned with providing a rank to the FP module on the basis of degree of fault-proneness. Fuzzy ordering algorithm is discussed in Sect. 5.4.4.

5.4 Model Implementation

The model is implemented in MATLAB and based on three major steps as

1. Selection of training data (preprocessing)
2. Decision tree construction (learning)
3. Module ranking

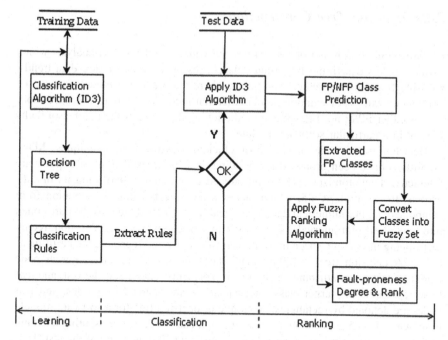

Fig. 5.1 Architecture of the proposed model

5.4.1 Training Data Selection

In order to provide better predictive results, the predictive model is trained with known data. For this, subset of data set with known result are taken. Training data selection is the most vital part for any supervised learning algorithms. It has been observed that most of the real-world project data are noisy, missing, and redundant due to their size, complexity, and various sources from where they are derived and collected. These data must be preprocessed to get high-quality training data.

Incomplete, noisy, and redundant data are common place properties of several real-world project data. There are many possible reasons for these anomalies. Therefore, data must be preprocessed before using it. There are a number of data preprocessing techniques such as data cleaning, integration, transformation, and reduction. Data cleaning can be applied to remove noise and correct inconsistencies in the data. Data integration merges data from multiple sources into an acceptable form. Data transformation, transform or consolidate the data into forms appropriate for training. Data reduction is applied to obtain a reduced representation of the dataset that is much smaller in volume, yet closely maintains the integrity of the original data.

5.4.2 Decision Tree Construction

Classification and prediction techniques are widely used for data analysis which extract models describing important data classes or to predict future data trends. Classification models predict class value, whereas a prediction model predicts the continuous value. For example, a classification model categorizes software modules as either FP or not FP, while a prediction model predicts the number of faults present in a particular software module.

Decision tree is one of the most efficient classification techniques. Many algorithms have been proposed for building decision trees. The most popular are Interactive Dichotomizer 3 (ID3) introduced by Quinlan (1986), and its modifications, for example, C4.5, which makes a decision tree for classification from symbolic data. For numerical data, numerical range of attribute must be partitioned into several disjoint intervals. These algorithms recursively create decision tree by partitioning the sample space in a data-driven manner and represent the partition as a tree. A decision tree is a flow-chart-like structure where each internal node denotes a test on an attribute, each branch represents outcome of the test, and leaf node represents the target class. Internal nodes are denoted by circles, and leaf nodes are denoted by rectangles. A sample of decision tree for a software module is shown in Fig. 5.2. As shown in the Fig. 5.2, each node is representing a testing attributes (software metrics), each branch represents the outcome of the test (low, medium, or high), and target classes are FP and NFP.

Decision trees are comprised of two major procedures: (1) building procedure or induction and (2) classification procedure or inference. Building procedure starts with an empty tree to select an appropriate test attribute for each decision node using some attribute selection measure. The process continues for each subdecision tree until reaching leaves with their corresponding class. The knowledge represented in decision tree can be extracted and represented in the form of classification "if–then" rules. The various classification rules may take the form as follows:

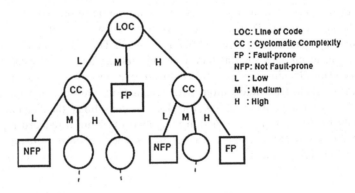

Fig. 5.2 A simple decision tree

- If LOC = L and CC = L then FP = "No."
- If LOC = M then FP = "Yes."
- If LOC = H and CC = L then FP = "No."
- If LOC = H and CC = H then FP = "Yes."

Attribute selection is generally based on information gain, which serves as a criterion in choosing test attribute at each decision node. Partitioning consists of partitioning the training set according to all possible attributes values, which leads to the generation of one partition for each possible value of the selected attribute. Stopping stops the partitioning process if (1) all the remaining objects belong to only one class, then the node is declared as a leaf labeled with this class value (2) there is no further attribute to test. For software module classification, ID3 algorithm steps can be summarized as follows:

(1) Take all the metrics of project data and count their information gain.
(2) Choose a metric as a test-metric with highest information gain value.
(3) Create nodes for this test-metric and do partition.
(4) Repeat (1)–(3) recursively until

   (i) All data points for a selected node belong to the same class.
   (ii) There are no more remaining metrics on which the data may be further partitioned.
   (iii) There are no data available for the selected node.

These steps can be applied to a set of data to generate decision tree that minimizes the expected value of the number of tests for classifying the data. When decision trees are built, many of the branches may reflect noise or noise in the training data. Tree pruning can be used to improve the classification accuracy by removing such irrelevant branches. Classification rules are extracted from the decision tree by tracing a path from the root to a leaf node. The accuracy of the derived classifier can be estimated using confusion matrix and discussed in Sect. 4.5. Furthermore, a classifier can classify software modules as FP or Not FP but it cannot assign the rank to a module on the basis of degree of fault-proneness. This module ranking can be done using fuzzy inference system and discussed in Sect. 5.4.4.

5.4.3 Estimating Classifier Accuracy

Training data are used in present study to derive the classifier, and then, different test datasets are applied to the classifier to estimate the accuracy of the model. Accuracy of the model is considered as the comparison factor with the earlier traditional models and may be obtained using confusion matrix as given in Table 5.1. A confusion matrix contains information about actual and predicted classifications done by a classification system. Performance of such systems is commonly evaluated using the data in the matrix. In general, a confusion matrix is

Table 5.1 A confusion matrix

Predicted			
	Label	FP	NFP
Actual	FP	True positive (tp)	False negative (fn)
	NFP	False positive (fp)	True negative (tn)

of size $L \times L$, where L is the number of different class label. The following table shows the confusion matrix for a two-class classifier.

- tp is the number of correct predictions that an instance is positive
- tn is the number of correct predictions that an instance is negative
- fp is the number of incorrect predictions that an instance is positive
- fn is the number of incorrect of predictions that an instance negative

Some authors such as, El-Emam et al. (2001), Elish and Elish (2008) have used confusion matrix as the basis for determining predictive accuracy of the classifier. The predictive accuracy, also known as correct classification rate, is defined as the ratio of the number of modules correctly predicted to the total number of modules. It is calculated as follows:

$$\text{Accuracy} = \left[\frac{(tp + tn)}{(tp + tn + fp + fn)} \right] \times 100$$

For example, consider a model which predicts for 10,000 software modules whether each module is FP or Not FP. This model correctly predicts 9,700 NFP and 100 FP. The model also incorrectly predicts 150 modules which are NFP to be FP and 50 modules which are FP to be not FP. Therefore, in this case, model accuracy is found to be 98%.

5.4.4 Module Ranking Procedure

Let M_1, M_2, ..., M_n are the FP software modules, m_1, m_2, ..., m_n are the values of various metrics of each module, and μ_1, μ_2, ..., μ_n are the membership value of metrics in the module. Each software module is considered as a fuzzy set whose elements are the software metrics with different membership values. Thus, each software module can be described using Zadeh's notation (Zadeh 1965) as:

$$\widetilde{M_1} = \left\{ \frac{\mu_1}{m_1} + \frac{\mu_2}{m_2} + \frac{\mu_3}{m_3} + \cdots + \frac{\mu_n}{m_n} \right\}$$

$$\widetilde{M_2} = \left\{ \frac{\mu_1}{m_1} + \frac{\mu_2}{m_2} + \frac{\mu_3}{m_3} + \cdots + \frac{\mu_n}{m_n} \right\}$$

$$\vdots$$

$$\widetilde{M_n} = \left\{ \frac{\mu_1}{m_1} + \frac{\mu_2}{m_2} + \frac{\mu_3}{m_3} + \cdots + \frac{\mu_n}{m_n} \right\}$$

Membership functions of a software metrics can be developed by selecting a suitable method from the various available methods such as triangular, trapezoidal, gamma, and rectangular. However, triangular membership functions (TMFs) are widely used for calculating and interpreting reliability data because they are simple and easy to understand (Yadav et al. 2003). Further, it is assumed that software metrics are of logarithmic nature and can be divided into three linguistic categories low (*L*), medium (*M*), and high (*H*), depending on their value. Considering these, fuzzy profile ranges of software metrics are developed as:

$$FPR = \left[1 - \frac{\{\log_{10}(1 \text{ to } 3)\}}{\{\log_{10}(3)\}}\right]$$

Thus, the fuzzy profile of each software metrics may take the values: low (0; 0; 0.37), medium (0; 0.37; 1.0), and high (0.37; 1.0; 1.0). Now, module $\tilde{M}_1$ will be greater than $\tilde{M}_2$ if the following condition satisfies,

$$T\left(\tilde{M}_1 \geq \tilde{M}_2\right) = \max_{(x_1 \geq x_2)} \left\{\min\left(\mu_{\tilde{M}_1}\left((x_1)\right), \mu_{\tilde{M}_2}(x_2)\right)\right\}$$

where, $T\left(\tilde{M}_1 \geq \tilde{M}_2\right)$ is the truth value on the interval [0, 1]; $\mu_{\tilde{M}_1}(x_1)$ and $\mu_{\tilde{M}_2}(x_2)$ represent the degree of membership of first element of module M1 and M2, respectively. For k software modules, the truth values of an ordinal ranking can be found as:

$$T\left(\tilde{M} \geq \tilde{M}_1, \tilde{M}_2, \ldots \tilde{M}_k\right) = \left[T\left(\tilde{M} \geq \tilde{M}_1\right) \wedge T\left(\tilde{M} \geq \tilde{M}_2\right) \wedge \ldots \wedge T\left(\tilde{M} \geq \tilde{M}_k\right)\right]$$

$$T\left(\tilde{M} \geq \tilde{M}_1, \tilde{M}_2, \ldots \tilde{M}_k\right) = \min\left[T\left(\tilde{M} \geq \tilde{M}_1\right), T\left(\tilde{M} \geq \tilde{M}_2\right), \ldots T\left(\tilde{M} \geq \tilde{M}_k\right)\right]$$

5.4.5 An Illustrative Example

Let there are three software modules $\tilde{M}_1$, $\tilde{M}_2$, and $\tilde{M}_3$ with only two known metrics, CC and DC for each module. These modules can be represented as a fuzzy set as:

$$\tilde{M}_1 = \left\{\frac{H}{m_1} + \frac{M}{m_2}\right\}, \tilde{M}_2 = \left\{\frac{M}{m_1} + \frac{H}{m_2}\right\} \text{ and } \tilde{M}_3 = \left\{\frac{M}{m_1} + \frac{H}{m_2}\right\}$$

After assigning the values to the membership function, we have

$$\tilde{M}_1 = \left\{\frac{1.0}{m_1} + \frac{0.8}{m_2}\right\}, \tilde{M}_2 = \left\{\frac{0.7}{m_1} + \frac{1.0}{m_2}\right\} \text{ and } \tilde{M}_3 = \left\{\frac{0.8}{m_1} + \frac{1.0}{m_2}\right\}$$

After assigning the actual metric value to each metric inside the module, we get:

$$\tilde{M}_1 = \left\{ \frac{H}{3} + \frac{M}{7} \right\}, \tilde{M}_2 = \left\{ \frac{H}{4} + \frac{M}{6} \right\} \text{ and } \tilde{M}_3 = \left\{ \frac{H}{2} + \frac{M}{4} \right\}$$

Now, the truth values of the inequality, $\tilde{M}1 \geq \tilde{M}2$ as follows:

$$T(\tilde{M}_1 \geq \tilde{M}_2) = \max_{(x_1 \geq x_2)} \left\{ \min\left(\mu_{\tilde{M}_1}(x_1), \mu_{\tilde{M}_2}(x_2) \right) \right\}$$
$$= \max\left\{ \min\left(\mu_{M_1}(7), \mu_{M_2}(4) \right), \min\left(\mu_{M_1}(7), \mu_{M_2}(6) \right) \right\}$$
$$= \max\{ \min(0.8, 0.7), \min(0.8, 1.0) \} = \max\{0.7, 0.8\} = 0.8$$

Similarly, the following can be obtained:

$$T(\tilde{M}_1 \geq \tilde{M}_3) = \max\{ \min(1.0, 0.8), \min(0.8, 0.8), \min(0.8, 1.0) \} = 0.8$$

$$T(\tilde{M}_2 \geq \tilde{M}_3) = \max\{ \min(0.7, 0.8), \min(0.7, 1.0), \min(1.0, 0.8), \min(1.0, 1.0) \}$$
$$= 1$$

$$T(\tilde{M}_2 \geq \tilde{M}_1) = \max\{ \min(0.7, 1.0), \min(1.0, 1.0) \} = 1.0$$

$$T(\tilde{M}_3 \geq \tilde{M}_2) = \max\{ \min(1.0, 0.7) \} = 0.7$$

$$T(\tilde{M}_3 \geq \tilde{M}_1) = \max\{ \min(1.0, 1.0) \} = 1.0$$

Now $T(\tilde{M}_1 \geq \tilde{M}_2, \tilde{M}_3)$ can found by comparing module $\tilde{M}_1$, with $\tilde{M}_2$ and $\tilde{M}_3$ as,

$$T(\tilde{M}_1 \geq \tilde{M}_2, \tilde{M}_3) = \left[T(\tilde{M}_1 \geq \tilde{M}_2) \wedge T(\tilde{M}_1 \geq \tilde{M}_3) \right]$$
$$= \min\left[T(\tilde{M}_1 \geq \tilde{M}_2), T(\tilde{M}_1 \geq \tilde{M}_3) \right] = 0.8$$

Similarly,

$$T(\tilde{M}_2 \geq \tilde{M}_1, \tilde{M}_3) = 1.0, \text{ and } T(\tilde{M}_3 \geq \tilde{M}_1, \tilde{M}_2) = 0.7$$

After comparisons, the first rank is assigned to module $\tilde{M}_2$ as its fault-proneness degree is found to be 1.0, module $\tilde{M}_1$ is ranked to second with 0.8 values, and finally, the third is assigned to module $\tilde{M}_3$.

5.5 Proposed Procedure

A stepwise procedure for software module prediction and ranking is given below.

Step 1 Select training data (software metrics with associated values).
Step 2 Construct a decision tree using ID3 algorithm and training data as:

Step 2.1 Identify the target class C {P: FP, N: NFP}.
Step 2.2 Create a node N;

Step 2.3 If all instances are of the same class C, create a leaf node with label C; and exit.

Step 2.4 If metric-list is empty, then create a node as a leaf node labeled with the most common class in the sample and exit.

Step 2.5 Select test-metric, the metric among the metric-list with highest information gain (i.e., attribute selection).

Step 2.6 Label node N with test-metric (splitting metric);
For each known value (say a_i) of test-metric, grow a branch from node N for the condition test-metric $= a_i$; (i.e., partitioning).
If there are no samples for the branch test-metric $= a_i$; then a leaf is created with majority class in samples.

Step 2.7 Return (decision tree)

Step 3 Extract the classification rules form the decision tree.

Step 4 Classify the target data into two class say FP and NFP.

Step 5 Find all FP modules and represent each module as a fuzzy set.

Step 6 Develop fuzzy profile of software module.

Step 7 Find the degree of fault-proneness of each module using module ranking procedures discussed in Sect. 5.4.4.

Step 8 Rank FP modules on the basis of its degree of fault-proneness.

5.6 Case Study

5.6.1 Dataset Used

This study makes the use of KC2 project data which is a public domain dataset and available through metric data program (MDP) at NASA (2004). The KC2 project, which was implemented in the C++, is the science data processing unit of a storage management system used for receiving and processing ground data for missions. The original KC2 dataset contains 522 program modules, of which 107 modules have one or more faults while remaining 415 modules are fault-free, that is, have no software faults. These data are shown in the Appendix C.

Each program module in the KC2 was characterized by 21 software metrics (5 different lines of code metrics, 3 McCabe metrics, 4 base Halstead metrics, 8 derived Halstead metrics, 1 branch count) and 1 target metric, which says whether a software module is FP or not. Out of these 21 software metrics, only 13 metrics (5 different lines of code metrics, 3 McCabe metrics, 4 base Halstead metrics, and 1 branch count) are used because the remaining 8 derived Halstead metrics do not contain any extra information for software fault prediction. These

Table 5.2 Software metrics considered for the study

Metrics	Information
LOC	McCabe's line count of code
EL	Executable LOC
CL	Comment LOC
BL	Blank LOC
CCL	Code and comment LOC
n1	No. of unique operators
n2	No. of unique operands
N1	Total no. of operators
N2	Total no. of operands
CC	McCabe's cyclomatic complexity
EC	McCabe's essential complexity
DC	McCabe's design complexity
BC	Branch count of flow graph

thirteen metrics are provided in Table 5.2. Definitions of these metrics are given below:

LOC measures: There are five different lines of code measure: line count of code (LOC), executable LOC (EL), comment LOC (CL), blank LOC (BL), and code and comment LOC (CCL). Lines of code are measured according to McCabe's line-counting conventions.

Halstead measures: Halstead complexity measures are software metrics introduced by Maurice Howard Halstead in 1977. According to Halstead, metrics of the software should reflect the implementation or expression of algorithms in different languages, but be independent of their execution on a specific platform. These metrics are therefore computed statically from the source code. Halstead's goal was to identify measurable properties of software, and the relations between them. There are four base Halstead measures: number of unique operators (n1), number of unique operands (n2), total number of operators (N1), and total number of operands (N2).

CC: CC measures the number of "linearly independent paths." A set of paths is said to be linearly independent if no path in the set is a linear combination of any other paths in the set through a program's "flow graph." A flow graph is a directed graph where each node corresponds to a program statement, and each arc indicates the flow of control from one statement to another. CC is calculated by $CC = e - n + 2$, where "e" is the number of arcs in the flow graph, and "n" is the number of nodes in the flow graph.

EC: EC is the extent to which a flow graph can be "reduced" by decomposing all the subflow graphs of "G" that are "D-structured primes." Such "D-structured primes" are also sometimes referred to as proper one-entry one-exit subflow graphs. EC is calculated using $EC = CC - m$, where m is the number of subflow graphs of "G" that are D-structured primes.

DC: DC is the CC of a module's reduced flow graph. The flow graph, "G", of a module is reduced to eliminate any complexity which does not influence the interrelationship between design modules. According to McCabe, this complexity measurement reflects the modules calling patterns to its immediate subordinate modules.

BC: BC of the flow graph.

5.6.2 Converting Data in Appropriate Form

Original KC2 dataset are given on a numerical value (crisp value) against each metrics (as shown in the Appendix C). In order to use these datasets inside fuzzy inference system (FIS), data are to be converted into fuzzy range, that is, data values are to be fuzzified using expert opinion. As discussed in the previous section, only thirteen metrics are considered here for decision tree construction. These metrics values are fuzzified into three fuzzy value low (L), medium (M), and high (H) using expert opinion as shown in Table 5.3.

For the sake simplicity, only fourteen project data form KC2 dataset are taken just to explain the data conversion procedure. These datasets are given in Table 5.4. Table 5.5 shows data after converting them into low, medium, and high values using Table 5.3. Out of 22 metrics, here, we have taken only 5 metrics

Table 5.3 Expert-specified data conversion values

	Low	Medium	High
Size	≤10	11–50	≥51
CC	≤5	6–11	≥12
DC	≤2	3–5	≥6
N	≤10	11–100	≥101

Table 5.4 Fourteen modules data form KC2 dataset

MID	Size	CC	DC	N	Class
1	1.1	1.4	1.4	1.3	No
2	8	1	1	10	No
3	55	4	3	131	Yes
4	210	5	3	687	Yes
5	34	6	6	80	Yes
6	46	4	4	112	No
7	42	3	3	102	Yes
8	16	4	2	51	No
9	102	21	10	294	Yes
10	67	5	5	213	Yes
11	52	7	4	133	Yes
12	46	7	4	133	Yes
13	80	8	6	203	Yes
14	4	1	1	4	No

Table 5.5 Fourteen modules data after conversion

MID	Size	CC	DC	N	Class
1	L	H	L	L	No
2	L	H	L	L	No
3	M	H	M	L	Yes
4	H	M	M	H	Yes
5	H	L	H	M	Yes
6	H	L	M	H	No
7	M	L	M	H	Yes
8	L	M	L	M	No
9	L	L	H	H	Yes
10	H	L	M	H	Yes
11	L	M	M	H	Yes
12	M	M	L	H	Yes
13	M	H	H	H	Yes
14	H	M	L	L	No

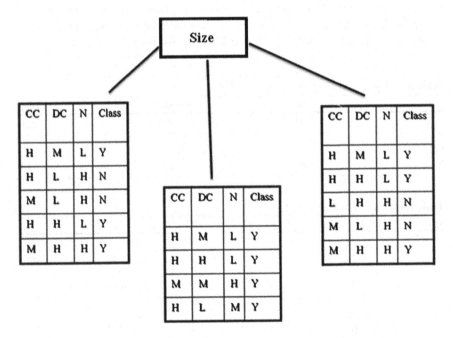

Fig. 5.3 An intermediate step of decision tree generation

(CC, DC, *N*: total no. of operators and operands, and Class: FP), just to illustrate the procedures. The decision tree is constructed using ID3 algorithm and classification rules are derived as given below (Figs. 5.3 and 5.4).

Thus, the classification rules are as follows:

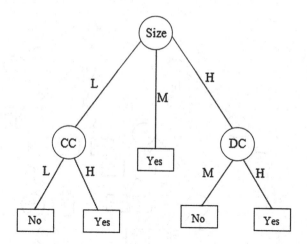

Fig. 5.4 A decision tree using fourteen datasets

1. If Size = *L* and CC = *L* then FP = "No."
2. If Size = *L* and CC = *H* then FP = "Yes."
3. If Size = *M* then FP = "Yes."
4. If Size = *H* and DC = *H* then FP = "Yes."
5. If Size = *H* and DC = *M* then FP = "No."

5.6.3 Resultant Decision Tree

Once data are converted, ID3 algorithm can be applied to construct decision tree. A MATLAB program is formulated for implementing ID3 algorithm which takes KC2 dataset and results a decision tree. Twenty, forty, sixty, and eighty percentage of KC2 datasets are used to generate different decision trees. Figures 5.5, 5.6, and 5.7 show the decision trees constructed using 20, 40, and 60% of KC2 datasets. After constructing, decision tree classification rules are derived as discussed in Sect. 5.4.2. These rules can be used as fuzzy rules with FIS to predict the FP and NFP modules. Prediction accuracy can be estimated using confusion matrix as discussed in Sect. 5.4.3. Once modules ranked on the basis of their degree of fault-proneness using the FIS and the procedure discussed in Sect. 5.4.4.

5.7 Results and Discussion

A program using MATLAB is formulated using the proposed approach using KC2 dataset (NASA 2004). Twenty, forty, sixty, and eighty percentage data of KC2 are used to derive the different decision tree (classifier), and then, accuracy of each

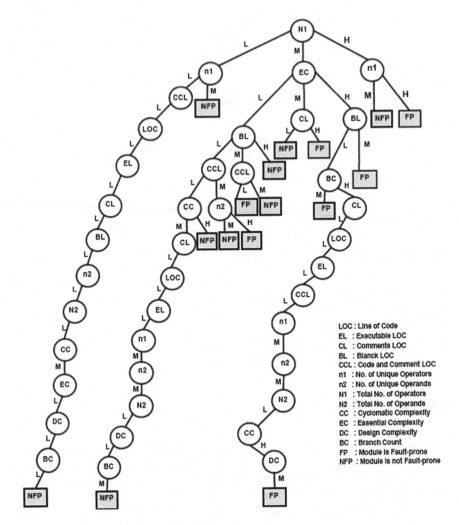

Fig. 5.5 Decision tree using 20 % of KC2 dataset

classifier is estimated on different mutually exclusive test data as shown in Table 5.6. The experiment is repeated ten times and each experiment type has been chosen as "Train/Test Percentage" of the data. Table 5.7 indicates that the prediction accuracy increases with increasing training data.

Model accuracy is estimated as the overall average accuracy obtained from ten different experiment results. The accuracy of the proposed model for KC2 dataset is presented in Table 5.6, which is found to better than some earlier models (Catal and Diri 2008; Kumar 2009) as shown in Table 5.9.

The model also predicts the ranks of the FP module on the basis of its degree of fault-proneness. Fuzzy ordering method is used for module ranking while fault-

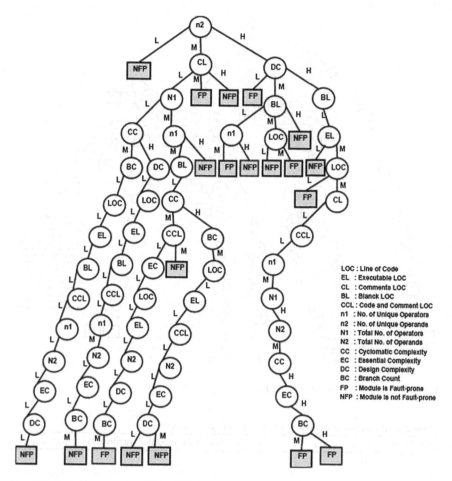

Fig. 5.6 Decision tree using 40% of KC2 dataset

proneness degree can be obtained using various FISs. Table 5.8 provides the ordered listing of various FP modules for each experiment as mentioned in Table 5.6. Model results are compared with two other existing models as shown in Table 5.9. Next, we show the effect of training on prediction accuracy. For this, we have developed six different MATLAB program, namely MP5_95, MP10_90, MP20_80, MP40_60, MP60_40, and MP20_80. Initially, when training data size is very small (5%), prediction accuracy is found to be 70.22% and it grows to 73.47% as soon as size of training data is increased to 10 percentage. It is found that on further increasing the size of training data, the prediction accuracy grows and reaches up to 95.08% as shown in Table 5.6.

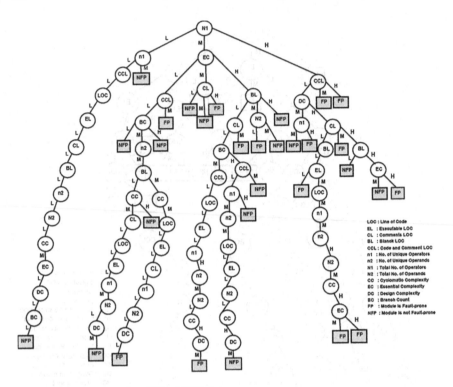

Fig. 5.7 Decision tree using 60% of KC2 dataset

Table 5.6 Prediction accuracy of the model for KC2 dataset

Experiment	Training (%)	Test (%)	Number of FP modules (predicted)	Accuracy (%)
1	20	20	17	81.73
2		40	42	78.85
3		60	54	80.13
4		80	58	75.37
5	40	20	6	84.62
6		40	19	87.50
7		60	26	86.02
8	60	20	25	85.58
9		40	47	88.74
10	80	20	10	95.08
			Average prediction accuracy	84.36

Table 5.7 Effect of Training on prediction accuracy

	MP5_95	MP10_90	MP20_80	MP40_60	MP60_40	MP80_20
Training (%)	5	10	20	40	60	80
Testing (%)	95	90	80	60	40	20
Accuracy (%)	70.22	73.47	75.37	86.02	86.84	95.08

Table 5.8 FP modules order in KC2 dataset

Experiment	Module ID (fault-proneness degree)
"20–20" (17 modules)	21 (0.5667), 14 (0.5214), 55 (0.5114), 9 (0.4603), 32 (0.3626), 13 (0.3626), 17 (0.3626), 25 (0.3438), 12 (0.2756), 41 (0.2742), 45 (0.2742), 49 (0.2412), 77 (0.2135), 55 (0.2135), 64 (0.2135), 71 (0.2135), 19 (0.2135)
"20–40" (42 modules)	1 (0.5667), 12 (0.5667), 19 (0.5667), 5 (0.5484), 18 (0.5484), 81 (0.5004), 16 (0.5004),13 (0.4603), 60 (0.4603), 15 (0.3453), 112 (0.3453), 98 (0.3453), 97 (0.3453), 94 (0.3438), 81 (0.3438), 76 (0.3438), 73 (0.3438), 70 (0.3438), 26 (0.3438), 24 (0.2742), 192 (0.2742), 187 (0.2742), 179 (0.2742), 178 (0.2275), 172 (0.2275), 166 (0.2156), 164 (0.2156), 162 (0.2156), 161 (0.2156), 158 (0.2025), 156 (0.2025), 105 (0.2025), 100 (0.2025), 92 (0.2025), 91 (0.2025), 85 (0.2025), 79 (0.2025), 77 (0.2025), 75 (0.2025), 74 (0.2025), 71 (0.2025), 69 (0.2025)
"20–60" (54 modules)	285 (0.5667), 283 (0.5667), 280 (0.5667), 255 (0.5667), 254 (0.5667), 251 (0.5667), 228 (0.5667), 227 (0.5484), 224 (0.5484), 178 (0.5484), 177 (0.5264), 174 (0.5264),161 (0.5264), 156 (0.5004), 153 (0.4603), 150 (0.4603), 112 (0.3453), 98 (0.3453), 97 (0.3453), 94 (0.3453), 81 (0.3453), 76 (0.3453), 73 (0.3453), 70 (0.3438), 287 (0.3438), 284 (0.3438), 281 (0.3438), 279 (0.3438), 257 (0.3438), 249 (0.3438), 230 (0.2156), 222 (0.2156), 221 (0.2156), 180 (0.2156), 172 (0.2156), 171 (0.2156), 165 (0.2156), 159 (0.2025), 157 (0.2025), 155 (0.2025), 154 (0.2025), 151 (0.2025), 149 (0.2025), 105 (0.2025), 100 (0.2025), 92 (0.2025), 91 (0.2025), 85 (0.2025), 79 (0.2025), 77 (0.2025), 75 (0.2025), 74 (0.2025), 71 (0.2025), 69 (0.2025)
"20–80" (58 modules)	41 (0.5667), 49 (0.5667), 46 (0.5667), 36 (0.5667), 35 (0.5530), 34 (0.5530), 33 (0.5530), 32 (0.5530), 31 (0.5530), 30 (0.5484), 35 (0.5484), 34 (0.5484), 33 (0.5484), 37 (0.5484), 34 (0.5484), 37 (0.5484), 378 (0.5484), 377 (0.5484), 376 (0.5484), 375 (0.5484), 374 (0.5484), 373 (0.5484), 372 (0.5484), 371 (0.5484), 370 (0.5484), 369 (0.5484), 368 (0.5484), 367 (0.5484), 366 (0.5484), 363 (0.5264), 350 (0.4603), 345 (0.4603), 292 (0.4603), 289 (0.3438), 227 (0.3438), 410 (0.3438), 412 (0.3438), 404 (0.3438), 403 (0.2756), 394 (0.2756), 388 (0.2756), 383 (0.2756), 372 (0.2156), 369 (0.2156), 361 (0.2156), 360 (0.2156), 355 (0.2156), 353 (0.2156), 341 (0.2156), 293 (0.2156), 253 (0.2156), 244 (0.2156), 243 (0.2156), 242 (0.2109), 241 (0.2109), 240 (0.2109)
"40–20" (6 modules)	104 (0.5667), 95 (0.5667), 93 (0.3902), 87 (0.3626), 86 (0.2756), 12 (0.2756)
"40–40" (19 modules)	175 (0.5195), 172 (0.5195), 199 (0.5195), 197 (0.5195), 191 (0.3902), 190 (0.3902), 184 (0.3902), 180 (0.3902), 176 (0.3626), 174 (0.3626), 171 (0.3626), 170 (0.2756), 169 (0.2756), 168 (0.2756), 150 (0.2156), 120 (0.2156), 178 (0.2156), 173 (0.2156), 45 (0.2156)

(continued)

5 Prediction and Ranking of Fault-Prone Software Modules

Table 5.8 (continued)

Experiment	Module ID (fault-proneness degree)
"40–60" (26 modules)	271 (0.5195), 268 (0.5195), 313 (0.5195), 311 (0.5195), 303 (0.3902), 302 (0.3902), 299 (0.3902), 297 (0.3902), 290 (0.3626), 289 (0.3626), 283 (0.3626), 279 (0.3626), 272 (0.3626), 270 (0.2756), 267 (0.2756), 266 (0.2756), 265 (0.2156), 264 (0.2156), 222 (0.2156), 274 (0.2156), 269 (0.2156), 128 (0.2156), 68 (0.2156), 29 (0.2156), 24 (0.2156), 11 (0.2156)
"60–20" (25 modules)	2 (0.5667), 66 (0.5453), 62 (0.5205), 59 (0.5195), 56 (0.4603), 5 (0.4603), 3 (0.4603), 83 (0.3902), 73 (0.3693), 72 (0.3693), 64 (0.3693), 61 (0.3693), 57 (0.3693), 55 (0.3693), 6 (0.3693), 4 (0.3693), 71 (0.3693), 58 (0.3626), 98 (0.3626), 79 (0.2742), 76 (0.2742), 70 (0.2424), 68 (0.2156), 60 (0.2156), 69 (0.2156)
"60–40" (47 modules)	4 (0.3758), 6(0.3758), 11(0.3693), 14(0.3910), 19 (0.2756), 33 (0.2756), 65(0.2756), 81(0.2756), 82 (0.2756), 88 (0.2756), 89 (0.2756), 128 (0.3693), 133 (0.2756), 146 (0.5195), 160 (0.4313), 164 (0.3902), 165 (0.5195), 166 (0.3902), 167 (0.5195), 168 (0.5195), 169 (0.2948), 170 (0.2753), 171 (0.5195), 172 (0.2742), 173 (0.3758), 175 (0.3693), 176 (0.2753), 177 (0.4810), 180 (0.2753), 181 (0.3693), 183 (0.2756), 184 (0.2742), 185 (0.2742), 186 (0.2742), 187 (0.2742), 188 (0.2756), 189 (0.2756), 190 (0.2756), 192 (0.2756), 193 (0.3902), 195 (0.5195), 197 (0.2756), 198 (0.3902), 208 (0.3693), 218 (0.2756), 219 (0.2756), 220 (0.2756)
"80–20" (10 modules)	5, 4 (0.5667), 3 (0.5667), 7 (0.5437), 6 (0.4537), 80 (0.3815), 32 (0.3626), 27 (0.2742), 25 (0.2412), 112 (0.2156)

Table 5.9 Model comparison for KC2 dataset

Model	Class prediction	Rank prediction	Accuracy (%)
Catal and Diri (2008)	Yes	No	82.22
Kumar (2009)	Yes	No	81.72
Proposed model	Yes	Yes	84.36

5.8 Summary

Although, a lot of work has been carried out toward classifying software modules as FP and NFP, the ranking of FP modules with degree of fault-proneness is missing. This study has proposed a new model for prediction and ranking of FP module for a large software system. ID3 algorithm is used to classify software modules as FP or not FP. At the same time, fuzzy ordering algorithm is applied to rank FP modules on the basis of their degree of fault-proneness. The proposed work has an endeavor to take care of this gap in the existing models. Ranking of FP module along with classification found to be a new approach to help in prioritizing and allocating test resources to the respective software modules. The proposed model has been applied and compared with existing model for the KC2 data of NASA. The results observed are promising and exhibit good accuracy, when compared with some of the earlier models.

References

Catal, C., & Diri, B. (2008). A fault prediction model with limited fault data to improve test process. In *Proceedings of the 9th International Conference on Product Focused Software Process Improvement* (pp. 244–257).

El-Emam, K., Melo, W., & Machado, J. C. (2001). The prediction of faulty classes using object-oriented design metrics. *Journal of Systems and Software, 56*(1), 63–75.

Elish, K. O., & Elish, M. O. (2008). Predicting defect-prone software modules using support vector machines. *Journal of Systems and Software, 81*(2008), 649–660.

Fenton, N. (1991). *Software metrics: a rigorous approach.* London: Chapmann & Hall.

Fenton, N. E., & Neil, M. (1999). A critique of software defect prediction models. *IEEE Transaction on Software Engineering, 25*(5), 675–689.

Fenton, N. E., & Neil, M. (2000). Software metrics: roadmap. In *Proceedings of the Conference on The Future of Software Engineering* (pp. 375–370). Limerick, Ireland.

Han, J., & Kamber, M. (2001). *Data mining: concepts and techniques.* USA: Morgan Kaufmann Publishers.

Khoshgoftaar, T. M., & Allen, E. B. (1999). A comparative study of ordering and classification of fault-prone software modules. *Empirical Software Engineering, 4*, 159–186.

Khoshgoftaar, T. M., & Allen, E. B. (2003). Ordering fault–prone software modules. *Software Quality Journal, 11*, 19–37.

Khoshgoftaar, T. M., & Munson, J. C. (1990). Predicting software development errors using complexity metrics. *IEEE Journal on Selected Areas in Communication, 8*(2), 253–261.

Khoshgoftaar, T. M., & Seliya, N. (2003). Fault prediction modeling for software quality estimation: comparing commonly used techniques. *Empirical Software Engineering, 8*, 255–283.

Kumar, K. S. (2009). Early Software Reliability and Quality Prediction, Ph.D. Thesis, IIT Kharagpur, Kharagpur, India.

Kumar, K. S., Misra, R. B., & Goyal, N. K. (2008). Development of fuzzy software operational profile. *International Journal of Reliability, Quality and Safety Engineering, 15*(6), 581–597.

Musa, J. D., Iannino, A., Okumoto, K. (1987). *Software reliability: measurement, prediction, and application*. New York: McGraw–Hill Publication.

NASA (2004). NASA metrics data program, http://mdp.ivv.nasa.gov/.

Ohlsson, N., & Alberg, H. (1996). Predicting fault–prone software modules in telephone switches. *IEEE Transaction on Software Engineering, 22*(12), 886–894.

Quinlan, J. R. (1986). Induction of decision trees. *Machine Learning, 1*, 81–106.

Ross, T. J. (2005). *Fuzzy logic with engineering applications* (2nd ed.). India: Willy.

Schneidewind, N. F. (1992). Methodology for validating software metrics. *IEEE Transactions on Software Engineering, 18*(5), 410–422.

Yadav, O. P., Singh, N., Chinnam, R. B., & Goel, P. S. (2003). A fuzzy logic based approach to reliability improvement during product development. *Reliability Engineering and System Safety, 80*, 63–74.

Zadeh, L. A. (1965). *Fuzzy Sets: Information and Control, 8*(3), 338–353.

Chapter 6
Reliability Centric Test Case Prioritization

6.1 Introduction

Software systems require various changes, throughout their lifetime, based on its faults, changes of user requirements, changes of environments, and so forth. It is very important to ensure that these changes are incorporated properly without any adverse effect on the quality and reliability of the software. In general, changing the software to correct faults or add new functionality can cause existing functionality to depart, introducing new faults. Therefore, software is regularly retested after subsequent changes, in the form of regression testing.

Regression testing is one of the important software maintenance activities that allow software testing personnel to ensure the quality and reliability of modified program. Regression testing is necessary to validate the software after modification but it regarded as an expensive activity. To reduce the cost of regression testing, researches have proposed many techniques such as regression test selection (RTS), test-suite minimization (TSM), and test case prioritization (TCP). RTS techniques reduce the cost of regression testing by selecting an appropriate portion of the existing test-suite for use in revalidating the software system. TSM techniques (Harrold et al. 1993) cut the cost of regression testing by reducing the test-suites to a minimal subset that yields equivalent coverage with respect to some test adequacy criteria. Both RTS and TSM techniques reduce the cost of regression by compromising the fault detection capability and to scarify the quality and reliability of the system under test. TCP techniques reduce the cost of regression testing without affecting the fault detection capabilities.

TCP techniques neither reduce the test-suite size (TS) nor omit test cases. These techniques schedules all the test cases in the test-suite such that the test cases with highest priority, assigned through certain criterion, are executed earlier in the regression testing process. Several TCP techniques (Rothermel et al. 1999; Elbaum et al. 2000, 2002) presented to reduce the cost of regression testing. TCP itself is a time consuming and costly process to increase the test-suite's fault detection rate. For example, a test-suite of m test cases and a program of n statement, total required statement coverage prioritization required time is $O\ (m\ n\ +\ m\ \log\ m)$ and

A. K. Pandey and N. K. Goyal, *Early Software Reliability Prediction*,
Studies in Fuzziness and Soft Computing 303, DOI: 10.1007/978-81-322-1176-1_6,
© Springer India 2013

cost is $O(m\,n)$. More importantly, in some cases, the prioritization cost will grow to $O(m^2\,n)$, a factor of m more than total statement coverage prioritization (Rothermel et al. 1999; Elbaum et al. 2000, 2002). This motivates to develop a reliability centric cost-effective TCP technique which will provide a better fault detection rate.

In this chapter, we have discussed an integrated and cost-effective approach to test prioritization that increases the test-suite's fault detection rate. The proposed approach considers three important factors: program change level (PCL), test-suite change level (TCL), and test-suite size (TS), before application of any technique to execute test cases. These factors can be derived using the information from modified program version. Furthermore, a new test case prioritization approach based on two-step process: inter-group prioritization, and intra-group prioritization is developed and used.

The rest of the chapter is organized as follows. The next section of this chapter provides related work in this area. Section 6.3 formally defines TCP problem, its goals, measures, and technique in detail. Section 6.4 presents the conceptual framework of the proposed model. Section 6.5 discusses the results, and Sect. 6.6 summarizes the chapter with future scope.

6.2 Earlier Works

To reduce the cost of regression testing, Harrold et al. (1993) presented a new TSM technique to select a representative set of test cases from a test-suite that provides the same coverage as the entire test-suite. This selection is performed by identifying, and then eliminating the redundant and obsolete test cases in the test-suite. Rothermel and Harrold (1996) have presented a framework for evaluating RTS techniques that classifies techniques in terms of inclusiveness, precision, efficiency, and generality. They also suggested a criteria of safe and unsafe selection of regression test according to the situation. Some other techniques for RTS can also be found in the study of Rothermel and Harrold (2006) and Wong et al. (1997). In 1997, Wong et al. have found that both TSM and TCP suffer from certain drawback in some situation and suggested TCP according to the criterion of increasing cost per additional coverage. In 1999, Rothermel et al. described several techniques for prioritizing test cases and they empirically evaluated their ability to improve rate of fault detection—a measure of how quickly faults are detected within the testing process. For this, they provided a metric APFD, which measures the average cumulative percentage of faults detected over the course of executing the test cases in a test-suite in a given order. They showed that TCP can significantly improve the rate of fault detection of test-suites.

Elbaum et al. (2000) had empirically examined the abilities of several TCP techniques to improve the rate of fault detection of test-suites. They focused on version-specific TCP and demonstrated that version-specific TCP techniques can improve the rate of fault detection of test-suites in regression testing.

In 2001, Rothermel et al. formally defined the TCP problem, presented several techniques for prioritizing test cases, and presented the results of empirical studies in which those techniques were applied to various programs. Later, Elbaum et al. (2002) extended the work presented by Rothermel et al. (2001). They have considered "version-specific prioritization" and coarse granularity techniques which is not considered by Rothermel et al. (2001). In 2003, Elbaum et al. presented a work closely related to. They urged that neither of previous studies examined how the type and magnitude of changes affect the cost-effectiveness of RTS techniques nor consider various change characteristics with respect to TCP. Considering these facts, Elbaum et al. (2004) discussed a cost-effective TCP technique.

Do et al. (2006) replicated the study of TCP techniques, focusing on an object-oriented language, Java, that is, rapidly gaining usage in the software industry. Qu et al. (2007) proposed a general process of TCP for regression testing in black box environment. The main idea of these algorithms is to group all reused test cases by the fault types revealed and then dynamically adjust priorities according to test results. Park et al. (2008) presented a historical value-based approach, which is based on the use of historical value to estimate the current cost and fault severity for a cost-cognizant TCP technique. In 2009, Khan et al. examined the impact of test case reduction and prioritization on software testing effectiveness using previous studies such as Rothermel et al and Elbaum et al. (2002, 2004). They had discussed the practical applicability of reduction and prioritization techniques and present useful combinations appropriate for different testing environments. Fazlalizadeh et al. (2009) presented a new equation to compute the priority of test cases in each session of regression testing. The equation considers the environment, time and resource constraints. In 2010, Kim and Baik presented an effective fault aware TCP by incorporating a fault localization technique. In their work, they described an overview of TCP and fault localization.

Reviews of literature indicate that earlier test case prioritization techniques did not considered the factors such as, PCL, test-suite change level (TCL), and TS that affect the cost-effectiveness of the prioritization techniques. Also, all the traditional prioritization techniques have used a straightforward approach using some coverage criteria. These traditional techniques are found to be based on coverage information, which relies on data gathered on the original version of a program (prior to modifications). They have ignored the information from the modified program version that affect the cost of prioritization. Keeping these points in mind, a reliability centric, cost-effective test case prioritization approach is proposed in this chapter.

6.3 Test Case Prioritization

TCP techniques schedule test cases execution order according to some criterion (Rothermel et al. 1999). The criteria may be to schedule test cases in an order that achieves earliest code coverage, or exercises features in order of expected frequency to use, or exercise subsystems in an order to find the desired behavior. The prioritization problem was first introduced by Wong et al. (1997) as a flexible method of software regression testing. They first selected test cases based on modified code coverage and then prioritized them. Research in this context has been followed by Rothermel et al., Elbaum et al. (2000, 2002, 2003, 2004) and other researchers which resulted in a widely accepted formal definition of the problem with various TCP techniques. Formally, TCP problem can be defined as follows:

Given, T, a test-suite; PT, the set of permutations of T; f, a function from PT to the real numbers.

Find $T' \in PT$ such that $(\forall T'')(T'' \in PT)(T'' \neq T')[f(T') \geq f(T'')]$.

Here, PT represents the set of all possible prioritizations (orderings) of T, and f is a function that, applied to any such ordering, yields an award value for that ordering. TCP can address a wide variety of goals, including the followings:

- Testers may wish to increase the test-suite's fault detection rate, that is, the likelihood of revealing faults earlier during regression tests (early detection yield early fixing of faults for reliability growth).
- Testers may wish to increase the coverage of code in the system under test at a faster rate, allowing a code coverage criterion to be met earlier in the test process.
- Testers may wish to increase their confidence in the reliability of the system under test at a faster rate.

Among various stated goal of TCP, one potential goal is to increase a test-suite's rate of fault detection—that is, how quickly test-suite detects faults during the testing process. An increased rate of fault detection can provide earlier feedback on the system under regression test, allowing earlier debugging, and supporting faster strategic decisions. This chapter has focused on increasing the likelihood of revealing faults earlier in the testing process.

6.3.1 Test Case Prioritization Techniques

Given any prioritization goal, various TCP techniques may be utilized to meet the goal. For example, to increase the fault detection of test-suites, test cases may be prioritized in terms of their failure tendency as observed in the past. Alternatively, they may be prioritized in term of their increasing cost per-coverage of code or increasing cost per-coverage of features listed in requirement specifications. In any case, the intent behind the choice of a prioritization technique is to increase the likelihood of the prioritized test suite meeting the goal compared to an adhoc or random ordering of test cases. This chapter focuses on the first goal listed above: increasing the likelihood of revealing faults earlier in the testing process. From the various literature reviews, around 20 different TCP techniques are found. Details on various TCP techniques can be found in (Harrold et al. 1993; Rothermel et al. 1999; Elbaum et al. 2000, 2002).

6.3.2 APFD Metric

One way of assessing performance of a prioritized test-suite is its fault detection rate. Test-suite's detection rate is a measure which shows how quickly these known faults are detected. A fast detecting prioritized test-suite exhibits earlier fault detection than a slow detecting prioritized test-suite. As a measure of how rapidly a prioritized test-suite detects faults, average percentage of faults detected (APFD) metric is used. This metric measures the weighted average of percentage of faults detected over the life of a test-suite (Rothermel et al. 1999). APFD value ranges from 0 to 100, and higher number implies faster (better) fault detection rates. Let T be a test-suite containing n test cases, and let F be a set of m faults revealed by T. Let TFi be the position of first test case in T' of T that reveals fault i. The APFD for test-suite T' can be found using the equation as given below.

$$APFD = 1 - \frac{TF_1 + TF_2 + \cdots + TF_m}{nm} + \frac{1}{2n}.$$

Consider a program with 10 faults and a set suite of five test cases, A to E, with fault detecting abilities, as shown in Table 6.1. Suppose that tester has created an initial test-suite T in order A–B–C–D–E to test the program. This test-suite can be used to test the program by scheduling the test cases in various orders. Here, three order (techniques): unordered, random, and optimal are considered to test the program and evaluate their effectiveness. Unordered (untreated) techniques schedules test cases in the order they were constructed initially. Let test cases A, B, C, D, and E constructed earlier are ordered as A-B-C-D-E, say it test-suite T1. Similarly, test cases can be scheduled randomly in the order E–D–C–B–A say it T2, and optimally whose test case ordering is C–E–B–A–D say it T3. Now, the effectiveness of these three prioritization techniques can be assessed using APFD measure as explained in the following example.

An Illustrative example Let the test cases are executed in the order A–B–C–D–E. Table 6.2 shows the fault detection ability of the test-suite T1. Figure 6.1 shows the percentage of detected faults versus the fraction of the test-suite T1 used. The area under the curve thus represents the weighted average of the percentage of faults detected over the life of the test-suite. This area is the prioritized test-suite's APFD and found to be 50% for unordered TCP (T1).

Table 6.1 Fault exposed by test-suite

Test case	Fault									
	1	2	3	4	5	6	7	8	9	10
A	x				x					
B	x				x	x	x			
C	x	x	x	x	x	x	x			
D					x					
E								x	x	x

Table 6.2 Fault detection ability of T1

Test case	Fault									
	1	2	3	4	5	6	7	8	9	10
A	*				*					
B	*				*	*	*			
C	*	*	*	*	*	*	*			
D					*					
E								*	*	*
TF_i	1	3	3	3	1	2	2	5	5	5

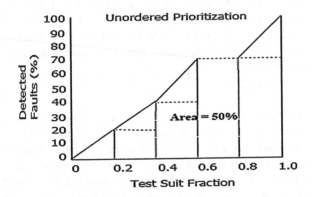

Fig. 6.1 APFD calculation for T1

APFD value can be calculated by putting the value of m, n, and TFi (the positions of the first test case in T that expose fault i) in APFD equation as follows:

$$APFD = 1 - \frac{1+3+3+3+1+2+2+5+5+5}{5 \times 10} + \frac{1}{2 \times 5} = 0.50.$$

Similarly, if test cases are executed in the random order say it $T2$ [E–D–C–B–A]. The APFD value is found to be 64% which indicates that random prioritization is faster than unordered prioritization. If test cases are executed in the optimal order C–E–B–A–D, APFD can be found 84% which is the best value for the given situation. Therefore, the optimal approach APFD value are used for comparision with the proposed approach.

6.4 Proposed Model

Cost-effectiveness of prioritization techniques varies with several factors, including the characteristics of the software under test, the attribute of test cases used in testing, and modifications made to create the new version of the system. Taking this into consideration, an integrated, cost-effective TCP model is proposed. Model architecture is shown in Fig. 6.2.

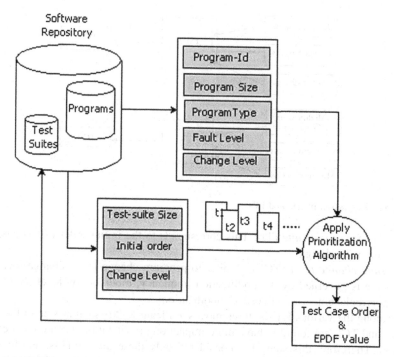

Fig. 6.2 Model architecture

The proposed model incorporates earlier information about software program and test-suite to determine the priority of test cases. Software programs are stored in the repository along with various related information such as program ID, program size, program type, fault level, and change level. Similarly, test-suites are also stored in the repository with information such as TS, initial order, and change level. These information have an significant impact on regression testing and can be referred as regression test metrics (RTMs). Some identified RTMs are as follows:

Program ID (PI): Every program in the repository is identified by its unique program ID.

Program size (PS): Program size corresponds to its size in terms of KLOC.

Program type (PT): The programs are categorized into two types: fault-prone (FP) or not fault-prone (NFP).

Program fault level (PFL): This measure can be derived using the information such as number of faults present in the program and severity of the faults.

Program change level (PCL): This measure is derived from the various measures related to modification of the program such as change frequency, change severity, cost etc. PCL value can be assigned three linguistic levels as low, medium, or high.

Test-suite size (TS): This corresponds to the number of test cases in a particular test-suite.

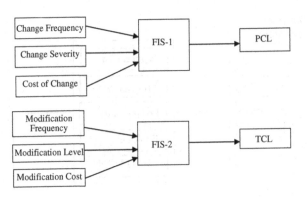

Fig. 6.3 Regression metric and FIS

Initial order of test-suite (TI): This corresponds to the initial order of test case at the time of its design.

Test-suite change level (TCL): Similar to program change level, change level of test-suite is governed by its modification frequency, modification level (addition/deletion/replacement), and cost of modification.

Among the various RTMs, three most significant RTMs such as size of the TS, PCL, and TCL are found to have more impact on prioritization when compared to others. Therefore, this study has considered only these three RTMs, and by utilizing the value of these metrics, a test case order is generated and APFD value is calculated.

The proposed approach focuses on the use of earlier information of the program and test cases to determine the value of regression metrics using fuzzy inference system (FIS). Fuzzy membership functions are generated utilizing the linguistic categories such as low (L), medium (M), and high (H) identified by an expert to express his/her assessment. The value of PCL, TCL, and TS are obtained using FIS as shown in Fig. 6.3:

The value of TS metrics is fuzzified into three fuzzy values: low (L), medium (M), and high (H), using application specific domain expert opinion as follows:

	Low	Medium	High
Test-suite size	≤2	3–6	≥6

A MATLAB program is formulated that takes the regression metrics value as input and results a test case order and APFD value as output. The proposed model integrates traditional TCP techniques along with a new TCP technique in a cost-effective manner. This new TCP approach is applied if the size of test-suite is high. For this, first, different groups of test cases are formed based on their similarity of detecting faults. After then, an intra-group prioritization followed be inter-group prioritization is performed to generate a complete test case order. Fig. 6.4 shows the steps involved in proposed TCP.

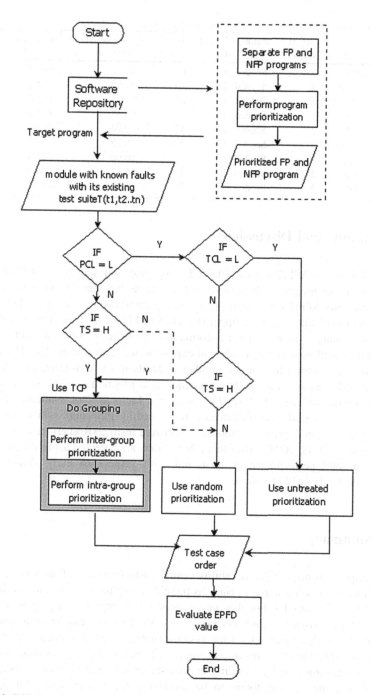

Fig. 6.4 TCP flowchart

Table 6.3 Program considered for the case studies with APFD values

Program	Size (LOC)	No. of version	Fault type	No. of fault	Test-suite size	Optimal APFD value (%)	Proposed APFD value (%)
P1	634	5	Seeded	8	6	94.12	88.44
P2	1246	4	Seeded	20	10	91.60	86.00
P3	409	5	Seeded	10	5	84.75	80.00
P4	792	4	Seeded	10	5	90.33	84.00
P5	1022	2	Seeded	15	12	80.00	74.90
P6	2158	2	Seeded	12	20	88.55	82.23
P7	4312	1	Seeded	10	16	80.00	72.00
P8	1293	2	Seeded	10	7	86.17	82.00
P9	1029	3	Seeded	8	7	79.00	74.22

6.5 Results and Discussion

Table 6.3 shows APFD values obtained using proposed approach with nine different in-house programs. It is obvious from Table 6.3 that the proposed prioritization results APFD value nearer to optimal prioritization as shown in Table 6.3. Optimal prioritization gives an upper bound of APFD value. This optimal ordering is possible only if there is prior information about the number of faults in the program as well as which test case can expose which fault if executed. Therefore, this study has seeded the known number of faults in each program to obtain an optimal order. In general, faults are created in a program version as a result of modifications producing that version. Such real faults are not available. To find such faults, this study has followed a fault seeding procedure which has been employed in several previous studies of software testing (Rothermel et al. 1999; Elbaum et al. 2002, 2004). After then, these faults are detected using the proposed technique and performance is measured with respect to early detection of regression faults (APFD value).

6.6 Summary

This chapter has presented a reliability centric cost-effective TCP approach. Focus is to detect regression faults as early as possible using the most significant RTMs, viz. PCL, TCL, and TS. For this, an integrated TCP approach is proposed which increases the test-suite's fault detection rate. An improved rate of fault detection can provide earlier feedback on the system, enabling earlier debugging, and thus increasing reliability. Moreover, the approach helps testing personnel to save regression testing cost by avoiding unnecessary prioritization. Model results are shown in Table 6.3 and found to be promising when compared with optimal prioritization techniques. Some useful directions for future work may include analyzing the impact of PCL, TCL, and TS on safety–critical software where certain faults can result in catastrophic consequences such as death, injury, or environmental harm. In those scenarios, prioritization should identify the critical faults early in order to make the system safe, risk free, and fail safe.

References

Do, H., Rothermel, G., & Kinneer, A. (2006). Prioritizing JUnit test cases: An empirical assessment and cost-benefits analysis. *Empirical Software Engineering, 11*, 33–70.

Elbaum, S., Malishevsky, A., & Rothermel, G. (2000). Prioritizing test cases for regression testing. In *Proceedings of International Symposium on Software Testing and Analysis* (pp. 102–112).

Elbaum, S., Malishevsky, A., & Rothermel, G. (2002). Test case prioritization: a family of empirical studies. *IEEE Transaction of Software Engineering, 28*(2), 159–182.

Elbaum, S., Kallakuri, P., Malishevsky, A., Rothermel, G., & Kanduri, S. (2003). Understanding the effects of changes on the cost-effectiveness of regression testing techniques. *Journal of Software, Verification and Reliability, 12*(2), 65–83.

Elbaum, S., Rothermel, G., Kanduri, S., & Malishevsky, A. G. (2004). Selecting a cost-effective test case prioritization technique. *Software Quality Journal, 12*(3), 185–210.

Fazlalizadeh, Y., Khalilian, A., Azgomi, M., & Parsa, S. (2009). Prioritizing test cases for resource constraint environments using historical test case performance data. In *Proceedings of IEEE conference* (pp. 190–195).

Harrold, M., Gupta, R., & Soffa, M. (1993). A methodology for controlling the size of a test suite. *ACM Transaction on Software Engineering and Methodology, 2*(3), 270–285.

Khan, S. R., Rehman, I., & Malik, S. (2009). The impact of test case reduction and prioritization on software testing effectiveness. In *Proceeding of International Conference on Emerging Technologies* (pp. 416–421).

Kim, S., & Baik J. (2010). An effective fault aware test case prioritization by incorporating a fault localization technique. In *Proceedings of ESEM-10* (pp. 16–17). Italy: Bolzano-Bozen.

Park, H., Ryu, H., & Baik, J. (2008). Historical value-based approach for cost-cognizant test case prioritization to improve the effectiveness of regression testing. In *Proceedings 2nd International Conference on Secure System Integration and Reliability Improvement* (pp. 39–46).

Qu, B., Nie, C., Xu, B., & Zhang, X. (2007). Test case prioritization for black box testing. In *Proceedings of 31st Annual International Computer Software and Applications Conference.*

Rothermel, G., & Harrold, M. J. (1996). Analyzing regression test selection techniques. *IEEE Transaction on Software Engineering, 22*(8), 529–551.

Rothermel, G., Untch, R. H., Chu, C., & Harrold, M. J. (1999). Test case prioritization: An empirical study. In *Proceedings of International Conference on Software Maintenance* (pp. 179–188).

Rothermel, G., Untch, R. H., Chu, C., & Harrold, M. J. (2001). Prioritizing test cases for regression testing. *IEEE Transaction on Software Engineering, 27*(10), 929–948.

Wong, W. E., Horgan, J. R., London, S., Agrawal, H. (1997). A study of effective regression testing in practice. In *Proceedings of 8th International Symposium on Software Reliability Engineering* (pp. 230–238).

Chapter 7
Software Reliability and Operational Profile

7.1 Introduction

Software testing (both development and regression) is costly and therefore must be conducted in a planned and systematic way to optimize overall testing objectives. One of the key objectives is to assure the reliability of software system that needs an extensive testing. To plan and guide software testing, the very first requirement is to find the number of tests (size) and allocation of these tests to various functions/modules. Traditional approach of deriving the test size is the function point (FP) approach by which various testing-related measures, such as the number of acceptance test cases, the total number of test cases, can be derived (Jones 1997). The limitation with the FP approach is that it is a static approach to estimate the size of a software system which is not well suited for embedded or real-time applications. For example, software consisting of four functions is developed for two classes of user. Class-I user will use certain functions more than class-II user and vice versa. FP approach is not able to allocate test cases; it simply counts the FP value and test cases. Furthermore, FP approach is not able to give the coverage information which is useful for reliability assurance of the system.

Reliability of a software-based product has bearing on how the computer and other external elements will use it (Musa et al. 1987; Musa 1993). A dynamic approach (dynamic test case derivation approach) estimates and allocates test cases as per the product field usages, which makes a good sense for reliability assurance. The approach is given by Musa (1993), called operational profile (OP). An OP provides the quantitative representation of how a system will be used in the real scenario by different users. It models how users use the different functions of a system with different occurrence probabilities. Such a description of the user behavior can be used to generate test cases and to guide testing to the mostly used functions. If developed early, an OP may be used to prioritize the development process, so that more resources are put on the most important operations. An OP improves the communication between customers and developers by blending their customer views on features they would like to have and their importance to them. In this way, the approach helps to achieve more reliable system, if tested

A. K. Pandey and N. K. Goyal, *Early Software Reliability Prediction*,
Studies in Fuzziness and Soft Computing 303, DOI: 10.1007/978-81-322-1176-1_7,
© Springer India 2013

accordingly. Limitation with OP approach is that it is useful when the estimates of test cases are available. The OP approach will not provide the size estimate about the test cases, rather it will guide to allocate these test cases in better way.

This chapter presents a reliability-centric operational profile-based testing (OPBT) approach by integrating FP and OP approaches. The proposed approach can be seen as an extension of Musa's model (1993). The approach can also be applied to traditional testing environment where it can be regarded as an enhanced traditional functional point testing model using the proposed functional complexity metric. Approach can also be used for test planning, automatic test case generation, and test case allocation.

The reminder of the chapter is organized as follows: Sect. 7.2 presents the background related to the current study. Section 7.3 describes the proposed model. Section 7.4 provides a case study on embedded electronic control unit (ECU). Section 7.5 covers the results and discussion. Section 7.6 summaries the chapter with the scope for the future work.

7.2 Backgrounds and Related Works

7.2.1 Embedded System Testing

Embedded ECU systems have many functional blocks such as power supply, processor, discrete/frequency/analog I/O. Testing these embedded ECUs is complex and time-consuming because of its dependency on the other ECUs, system, and subsystems. This challenge is further increased by the fact that the most of auto OEMs (Original Equipment Manufacturers) do not have a uniform testing process. It has been found that in many cases, the testing task is carried out by a vendor who may not have all the details about the system and its functional requirements.

7.2.2 Function Point Metric

Various approaches have been proposed for estimating the number of test cases such as ad-hoc approach, FP approach, and use case approach (Symons 1988; Fenton et al. 1998). Ad hoc testing is an informal approach to testing, with the goal of finding defects with every means possible. Tests are performed without much planning, that is, there are no written test cases. Use case approach identifies test cases that exercise the whole system on a transaction-by-transaction basis from start to finish from the user point of view.

Among the various available methodologies, FP is widely used for the purpose of test case derivation. Basically, FP consists of the weighted totals of five external aspects of software applications: the types of inputs, the types of outputs, the types of inquiries, which users can make, the types of logical files which the application maintains, and the types of interfaces to other applications (Symons 1988). FP approach has been in use for almost 20 years, and simple rules of thumb have been derived. Attempts have been made to establish a relationship between FPs and efforts associated with software development. Strong relationship exists between the number of defects, the number of test cases, and the number of FPs. The number of acceptance test cases can be estimated by multiplying the number of FPs by a factor of 1.2. Intuitively, the number of maximum potential defects is equal to the number of acceptance test cases, which is 1.2 x FPs (Jones 1997). Also, the total number of test cases can be estimated by FPs. Jones (1997) estimates that FPs raised to the power of 1.2 estimate the total number of test cases, that is, test cases grow at a faster rate than FPs.

$$\text{Number of acceptance test} = \text{Number of defects} = (1.2 \times \text{Function points})$$

$$\text{Total number of test cases} = (\text{Function points})^{1.2}$$

Limitation with the FP approach is that (1) it is not easy to apply in practice as it requires various calculation to derive the FP value. (2) it is not well suited for embedded or real-time applications because real-time functional measurement technique should account for subprocesses not only for the main process (Oligny et al. 1998b). Also, there is limited or no communication with external file or database. Therefore, the traditional FP approach needs to be revised as per the application requirements, and a realistic FP is to be derived. Earlier works in this area are as follows: full FP, feature point, 3D FP, application features, and Asset-R (Oligny et al. 1998a, b). Considering the limitations associated with the FP approach, a new FC (functional complexity) metric is advised for embedded systems. FC approach is an extension of FP approach designed for embedded systems, which is discussed in the later section of this book. The novelty of the proposed FC approach is the derivation of functional complexity from the specification documents.

7.2.3 Operational Profile

One of the pioneer researches about the development of OP is from John Musa from AT&T Bell Laboratories (Musa 1993). With an OP, a system can be tested more efficiently because testing can focus on the operations mostly used in the field.

It is a practical approach to ensure that a system is delivered with a maximum reliability, because the operations most frequently used also have been tested the most. Musa informally characterized the benefits-to-cost ratio as 10 or greater (Musa 2005). In 1993, AT&T had used an OP successfully for the testing of a telephone switching service, which significantly reduced the number of problems reported by customers (Koziolek 2005). Also, Hewlett-Packard reorganized its test processes with OPs and reduced testing time and cost for a multiprocessor

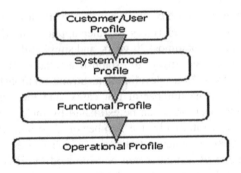

Fig. 7.1 Operational profile

operating system by 50% (Koziolek 2005). Arora et al. (2005) conducted a case study on Pocket PC, a Windows CE 3.0-based device and demonstrated a significant reliability improvement through OP-driven testing.

OP becomes particularly valuable in guiding test planning by assigning test cases to different operations in accordance with their probabilities of occurrence. OP development involves more than five steps (Musa 2005) as shown in Fig. 7.1.

(a) *Customer Profile* A customer type is one or more customers in a group who intend to use the system in a relatively similar manner and in a substantially different manner from other types of customer.

(b) *User Profile* A system's users may be different from the customers of a software product. A user is a person, group, or institution that operates, as opposed to acquires, the system. A user type is a set of users who will operate the system similarly.

(c) *System Mode Profile* A system mode is a way that a system can operate. The system includes both hardware and software. Most systems have more than one mode of operation. For each system mode, there may be more than one or two functional profiles. There are no technical limits on how many system modes may be established.

(d) *Functional Profile* This is the most important step during the OP development process. Once the system mode profile has been developed, functions performed during that mode can be identified and then probabilities are assigned to each of these functions. Functions are essentially tasks that an external entity such as a user can perform with the system. The functional profile can be either explicit or implicit, depending on the key input variables.

(e) *Operational Profile* After developing the functional profile, each function from each system mode is distributed into the number of operations and the occurrence probability is decided for each operation. An operation represents a task being accomplished by the system from the viewpoint of the people who test the system.

(f) *Test Cases Generation and Allocation* The decision on the total number of test cases is typically drawn based on the constraints of cost and time. Testing can also be guided with the help of statistical control charts, which allow one to know when testing can be stopped.

OP approach can give the size estimate of the test; it helps in test case allocation for better coverage and reliability assurance. Considering the two sensible constraints of FP approach and OP approach; an enhanced FC-OP-based approach is discussed in the next section.

7.3 Proposed Model

7.3.1 Premise

FC-based approach and OPBT approach can be used independently for testing. One can use the FC metric and further write test cases manually or using concepts such as orthogonal array testing. This offers a developer's view of testing and misses out the usage aspect. Important test cases may thus get missed out, though the coverage of specification can be achieved.

Optionally, one can use OPBT approach and do testing based on some assumed value for the total number of test cases. With respect to specification coverage, it is assumed that it will be taken care of, since all possible operations are considered. This offers a user's view of testing.

Proposal is to use FC metric to compute the total number of test cases and, at the same time, apply OPBT approach for allocation of test cases for efficient coverage and better reliability. This way, the developer's view can be combined with the user's view and an estimate of percentage coverage per function (in specification) can be obtained.

The next section presents a method for deriving relevant metrics to compute FC value, computing the number of test cases, and developing OPs and integrating them into *reliability-centric OPBT approach*. The method is applied, as a case study, to the front and rear fog light embedded ECU in automotive.

7.3.2 Model Architecture

The model architecture is shown in Fig. 7.2. A stepwise procedure is given below.

Step 1: Select the ECU with specifications.
Step 2: For enhanced traditional testing, go to step 3, and for reliability-driven testing, go to step 6.

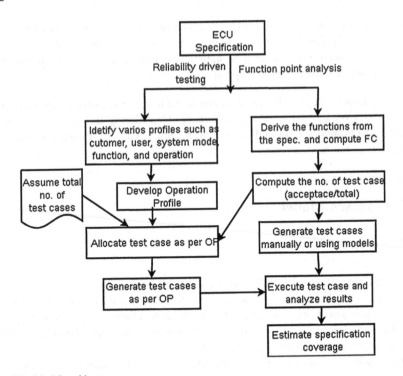

Fig. 7.2 Model architecture

Step 3: Derive the functions of ECU for the specification documents and functional complexity (FC) value.

Step 4: Compute the number of test cases.

Step 5: Generate test cases manually or using models and go to step 10.

Step 6: Develop the OP of the ECU.

Step 7: Assume total number of test cases based on the constraints of cost and time or use the number derived from FC analysis.

Step 8: Allocate test cases as per OP.

Step 9: Generate test cases as per OP.

Step 10: Execute test cases and analyze the results.

Step 11: Estimate specification coverage.

The proposed methodology integrates FC and OP to optimize the overall validation efforts with reliability assurance. FC metric can be arrived at considering FP and requirement complexity (RC) as below:

$$FC = 0.5 \times FP + 0.5 \times RC$$

FP metric is derived using weighted sum of input, output, file, interface, and inquiries as discussed in (Symons 1988; Fenton et al. 1998). RC measure represents the size and structural complexity of the requirements and is derived by counting the number of lines of non-trivial specification or by counting the number of decisions made in the specification (similar to cyclomatic complexity). On the basis of this FC metric, the required number of tests can be computed for each function as follows:

$$\text{Number of acceptance test case} = FC \times 1.2$$

Since FC is an extension of FP approach, the number of acceptance test cases can be estimated by multiplying the number of FPs by a factor of {1.2} and the total number of test cases can be obtained by FC raised to the power of 1.2. Once the total numbers of test cases are derived, allocation of these test cases to various functionalities is guided by the OP of the corresponding functions. OP-based test case allocation will improve the reliability since it focuses attention on how users will use the product with different probabilities. OP of the fog light ECU is developed, and test cases based on the OP are allocated.

7.4 Case Study: Automotive Embedded ECU

Automobile electronic systems can be broadly categorized into four main categories: 'power train drive' consisting of electronic engine management, electronic transmission, electronic networks; 'safety systems' such as antilock brake systems, air bag, antitheft, suspension, lighting, steering, skid systems; 'comfort body systems' such as air conditioner, seat adjusting, dashboard displays; 'communication systems' such as global positioning system, radio reception, information systems. Embedded ECUs are needed by these systems/subsystems for better performance.

Embedded ECUs, which utilize software to solve a very specific problem, are usually the most complicated and powerful computers in an automobile. Uses of software have become vital for the automotive industry. Current trends indicate that every year, there is a twofold increase in the size and complexity of software in high-end vehicles. Modern vehicles are equipped with approx 90–100 embedded ECU with varying degrees of complexity and size. Some ECUs are responsible for controlling critical function than others. Reliability and safety aspects of these ECUs are becoming challenging and should be verified and validated with respect to the specified functionality. Moreover, testing of system will never assure a complete coverage, and therefore, it is vital to find an optimal way of testing to assure the reliability of the product. To illustrate the proposed approach, a fog light ECU is considered and discussed in the next section.

7.4.1 Fog Light ECUs

The purpose of fog lights (front and rear) is to provide illumination for the driver to operate the vehicle safely during low visibility. Rear fog lights provide back side visibility to driver in order to avoid a rear-end collision as well. Major functions of fog light may include

1. Turn ON/OFF fog lights when ignition is ON.
2. Fog light control on the basis of other lights such as headlights, parking lights, and corner lights.
3. Status notification of above functionalities to the driver via instrument panel

These functions are distributed over two ECUs viz. body control module and instrument panel as shown in the following Fig. 7.3.

7.4.2 Functional Complexity of Fog Light ECU

Two major functionalities of front and rear fog light ECUs are identified from the specification documents. These functionalities are (1) front and rear fog light control and (2) key-on/off front and rear fog light. Fog light control functionality

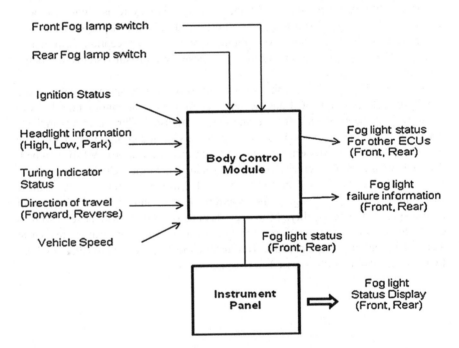

Fig. 7.3 Block diagram of front and rear fog light ECU

includes switching the front and rear fog lamps ON/OFF on the basis of certain specified conditions. To turn ON the front and rear fog lamps, the engine must be ON and either the park lamps or low beams must be active. Front and rear fog lights are activated by means of two independent front/rear momentary commands, originating in the headlamp switch assembly (HLS). Each push of the front and rear fog lamp switch toggles the front and rear off lamps ON to OFF, and OFF to ON. Depending on the ignition status (key-on/off), and the high-beam status, the front and rear fog lamps are temporarily turned off, and when the signal changes, the previous value of fog light is resumed. FP metric is derived using weighted sum of input, output, file, interface, and inquiries as discussed in Symons (1988).

The components are classified as simple, average, or complex, depending on the number of data elements in each type and other factors. The number of components and its classification are given in Tables 7.1 and 7.2, respectively. This study has assumed that all the components are of simple type and computed FP value as follows:

$$FP = 3 \times 4 + 4 \times 7 + 7 \times 3 + 5 \times 3 + 3 \times 0 = 76.$$

RC value using logical line count is found to be 58 for the current considered functionalities. Therefore,

$$FC = 0.5 \times FP + 0.5 \times RC = 0.5 \times 76 + 0.5 \times 58 = 67.$$

Now, the number of acceptance tests is computed as follows:

$$\text{Number of acceptance test} = (FC) \times 12 = 67 \times 1.2 = 80.4 \approx 80.$$

Table 7.1 Components of FP computation

User switches	External inputs	Value ranges	External outputs	User output	Internal logic files	External interface file
FFL switch FFL	FF command	On, Off	FFL (Left)	FL	active	FFL Int
RFL switch	RF command	On, Off	FFL (Right)	"	RFL Int	RFL
Ignition	Cmd Ign sts	Run, lock, start	RFL (Left)	"	FFL strategy	RFL fault status
HLS	HB, LB, park	0,1	RFL (Right)	"	–	–
– –	–	–	FFL display	Signal	mgmt.	–
–	–	–	RFL display	–	–	–
–	–	–	RFL fault display	–	–	–

FF front fog, *RF* rear fog, *FFL* front fog light, *RFL* rear fog light, *HB* high beam, *LB* low beam, *HLS* headlamp switch

Table 7.2 Components' classification

	Simple	Avg.	Complex
External input	3	4	5
External output	4	5	7
Internal logic files	7	10	15
External interface file	5	7	10
External inquiry	3	4	6

7.4.3 Operational Profile of Fog Light ECU

(a) *Development of customer profile, user profile, and system mode profile* Customer type and user type are same in the present case study, and hence, probability assigned is 1. For fog light ECUs, two system modes are identified as shown in Fig. 7.4.

(b) *Development of functional profile* The functional profile for running mode of vehicle has been developed and is shown in Fig. 7.5. On similar lines, functional profile for parked system mode can also be developed. The yearly uses of specific functions are indicative for a typical North American geography and are likely to change for different geographies as well as locales within North America.

(c) *Development of environmental variable profile* The functional profile can be either explicit or implicit, depending on the key input variables. Explicit profile consists of all enumerated functions and their associated probabilities of occurrence. Implicit profile consists of the set of values of key input variables, each with associated probability of occurrences. Environment variables' probability and profiles are listed in Tables 7.3 and 7.4, respectively.

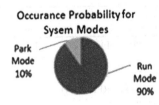

Fig. 7.4 Operating mode of front and rear fog light ECU

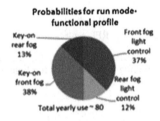

Fig. 7.5 Run mode functional profile of fog light ECU

Table 7.3 Environment variable probability

Environment variable	Status	Probability
Command ignition status	RUN	0.6
	START	0.2
	LOCK	0.2
Headlamp switch in part mode	ON	0.3
Headlamp switch in low-beam mode	ON	0.5
Headlamp switch in high-beam mode	ON	0.2
Front fog light command	ON	0.6
	OFF	0.4
Rear fog light command	ON	0.6
	OFF	0.4
Past status of front fog light	ON	0.5
	OFF	0.5
Past status of rear fog light	ON	0.5
	OFF	0.5

Table 7.4 Environment variable profile (F1)

#	Cmd Ign Sts	Prob.	Headlamp switch	Prob.	Fog lamp cmd	Prob.	Overall prob.
O1	RUN	0.6	HB	0.2	ON	0.6	0.072
O2	RUN	0.6	HB	0.2	OFF	0.4	0.048
O3	RUN	0.6	LB	0.5	ON	0.6	0.18
...	...	...	...	...	...	...	...

(d) *Development of operational profile* Each function from each system mode is distributed into number of operations. The occurrence probabilities are decided for each operation. The final OP is developed, combining all the operations for respective system modes. OP for running-mode vehicle is displayed in Table 7.5.

(e) *Test case allocation* The total number of test cases was decided on the basis of functional complexity. Therefore, the total number of test cases will be $= (FC)^{1.2} = (67)^{1.2} = 155$. Test cases to each operation were allocated on the basis of their OP as shown in the Table 7.5.

Table 7.5 Operational profile of running-mode vehicle

Fun.	Operation	Prob.	Overall probability	Test cases
F1	O1	0.072	0.0240	4
	O2	0.048	0.0160	2
	O3	0.18	0.060	9
F2	O1	...	...	...
F3	O1	...	...	...
F4	O1	...	...	...

7.4.4 Test Case Generation

Test cases are derived from the various possible values taken by different variables in each operation. This approach requires one to know all possible variables which can define a different states (or different operations) in a function. Variables such as ignition command and fog light command are accessible to the user, and it is possible to track the probability of these vairables. However, variables such as past status of front fog light are internal memory variables which are not accessible to the user. To avoid the use of such variables, the transition of states can be defined and its probability can be traced from the usage. The probability of initial states and the probability of transition can define all the possible operations from the user's perspective.

7.4.5 Test Cases and Transition Probability

Test cases are derived from the various possible states of front and rear fog light ECU. Totally thirty-six states are derived and are listed in Table 7.6. Transition probability from one state to another is also derived using expert opinion and similar/earlier project data, which is shown in Table 7.7. Many of these are zero, and using a matrix form can be cumbersome. Other forms such as Markov chains are being successfully used; however, for the present case study, a matrix form has been presented for simplicity. These values are then used for test planning and test case allocation.

7.5 Results and Discussion

Reliability-centric OPBT approach is presented in this chapter. For this, a case study is conducted for the front and rear fog light ECU. FP and FC values are derived, and total numbers of test cases are computed to be 181 and 155,

Table 7.6 States of front and rear fog light ECU

1	Ignition = RUN; low-beam headlamp = ON front fog command = ON; rear fog command = ON
2	Ignition = RUN; high-beam headlamp = ON front fog command = ON; rear fog command = ON
...	...
...	...
35	Ignition = LOCK; high-beam headlamp = ON; front fog command = ON; rear fog command = OFF
36	Ignition = LOCK; park light headlamp = ON; front fog command = ON; rear fog command = OFF

Table 7.7 Transition probabilities of states

	1	2	3	34	...	35	36
1	0	0.1	0.2	0	...	0	0
2	0.2	0	0.2	0	...	0	0
3	0.2	0.1	0	0	...	0	0
...	...	...	...	...	...	...	...
...	...	...	...	...	...	...	...
35	0	0	0	0.08	...	0	0.08
36	0	0	0	0.08	...	0.08	0

Note Only, sample of results are listed in the Tables 7.4, 7.5, 7.6, 7.7

Table 7.8 Comparisons of results

Approach	No. of test cases	Defect testing	Reliability testing	Cost (per/hrs)	Test case generation
Traditional FP	181	Yes	No	7.24	No
Traditional OP	Assumed	Yes	Yes	Depends on tests	Yes
Enhanced FC-OP Based	155	Yes	Yes	6.2	Yes

respectively. It is important to mention here that the proposed enhanced FC approach results in different (lesser) number of test cases as compared to the traditional FP approach as shown in Table 7.8. Also, an OP for the same ECU is developed and test cases are allocated to get the confidence in reliability. Moreover, the proposed OP-based approach can also be utilized to generate the test cases. The cost in terms of person/hour is found to be 6.2 for the enhanced FC-OP-based approach as compared 7.24 of the traditional FP-based approach. It has been assumed that 25 test cases are executed by a person in an eight-hour working day. Apart from addressing the uncertainty of reliability when traditional testing is done, the proposed approach also helps in test planning, test case generation, and coverage estimates. In future, the proposed approach can be applied to validate any automotive ECUs in a cost-effective way.

7.6 Summary

This chapter has presented a model-based approach to test embedded software by integrating the functional complexity and OP. The novelty of the proposed model is the derivation of functional complexity from the specification documents. An OPBT is also proposed to get a reliability estimate, reduce validation cost, and speedup V&V process by allocating the resources in relation to use and criticality. A case study of front and rear fog light ECUs is discussed. Only four functions are considered for the sake of simplicity, and test cases have been derived using the functional complexity. OP of each function has been utilized to allocate test cases and estimating coverage. The approach can be applied to other automotive ECUs for test planning and allocation of test cases. In future, the proposed approach can be extended to embedded ECUs of other domains such as medical devices, railways, avionics.

References

Arora, S., Misra, R. B., & Kumre, V. M. (2005). Software reliability improvement through operational profile driven testing. In *Proceedings of Annual IEEE Conference on Reliability and Maintainability Symposium* (pp. 621–627). Virginia.

Fenton, E., & Neil, M. (1998). *Software metrics: Successes, failures and new directions*. UK: Centre for Software Reliability, City University.

Jones, C. (1997). Software Productivity Research, Inc. Software Estimating Rules of Thumb, Version 1.

Koziolek, H. (2005). *Operational profiles for software reliability*. Germany: Seminar on Dependability Engineering.

Musa, J. D. (1993). Operational profiles in software reliability engineering. *IEEE Software Magazine*.

Musa, J. D. (2005). *Software reliability engineering: More reliable software faster and cheaper* (2nd ed.). New York: Tata McGraw-Hill Publication.

Musa, J. D., Iannino, A., & Okumoto, K. (1987). *Software reliability: Measurement, prediction, and application*. New York: McGraw-Hill Publication.

Oligny, S., Desharnais, J., & Abran, A. (1998a). Functional size of real time software: Overview of field test. UAQM's Software Engineering Management Research Laboratory.

Oligny, S., Desharnais, J., & Abran, A. (1998b). A method for measuring the functional size of embedded software. UAQM's Software Engineering Management Research Laboratory.

Symons, C. R. (1988). Function point analysis: Difficulties and improvements. *IEEE Transactions on Software Engineering, 14*(1), 2–11.

Appendix A

This appendix presents the list of metrics given by Li et al. (2003) along with their rank and fuzzy profile. The fuzzy range of the metrics is assigned using the document IEEE STD 982.2 and NASA (2004).

Table A.1 Metrics at requirements phase with fuzzy profiles

#	Metrics	Rank	Fuzzy range	Low	Moderate	High
1.	Fault density	0.71	[0–1]	(0;0.2;0.4)	(0.2;0.4;0.7)	(0.5;0.7;1;1)
2.	Requirement specification change requests	0.70	[0–100]	(0;0;35)	(25;45;75)	(60;100;100)
3.	Error distribution	0.68	[0–10]	(0;0;4.5)	(3;5.5;6.5)	(5.5;10;10)
4.	Reviews, inspection, and walkthroughs	0.61	[0–5]	(0;2;2)	(1;2;3)	(2;5;5)
5.	Software capability maturity level	0.60	[0–5]	(0;2;2)	(1;2;3)	(2;5;5)
6.	Fault-days number	0.60	[0–50]	(0;0;5)	(2;7;15)	(10;50;50)
7.	Function point analysis	0.51	[0–100]	(0;0;10)	(10;20;30)	(25;100;100)
8.	Requirements compliance	0.50	[0–100]	(0;2;5)	(3;10;20)	(20;100;100)
9.	Feature point analysis	0.46	[0–100]	(0;0;10)	(10;20;30)	(25;100;100)
10.	Number of faults remaining (error seeding)	0.46	[0–100]	(0;0;5)	(5;10;25)	(20;100;100)
11.	Cause and effect graphing	0.45	[0–1]	(0;0;0.2)	(0.2;0.4;0.7)	(0.5;1;1)
12.	Completeness	0.42	[0–1]	(0;0;0.4)	(0.2;0.5;0.7)	(0.5;1;1)

A. K. Pandey and N. K. Goyal, *Early Software Reliability Prediction*,
Studies in Fuzziness and Soft Computing 303, DOI: 10.1007/978-81-322-1176-1,
© Springer India 2013

Table A.2 Metrics at design phase with fuzzy profiles

#	Metrics	Rank	Fuzzy range	Low	Moderate	High
1.	Design defect density	0.75	[0–5]	(0;0;2)	(1;2;3)	(2;5;5)
2.	Fault density	0.73	[0–5]	(0;0;2)	(1;2;3)	(2;5;5)
3.	Cyclomatic complexity	0.73	[0–500]	(0;0;150)	(100;200;300)	(250;500;500)
4.	Fault-days number	0.71	[0–50]	(0;0;5)	(2;7;15)	(10;50;50)
5.	Requirement specification change requests	0.69	[0–100]	(0;0;35)	(25;45;75)	(60;100;100)
6.	Error distribution	0.68	[0–10]	(0;0;4.5)	(3;5.5;6.5)	(5.5;10;10)
7.	Man hours per major defect detected	0.63	[0–5]	(0;0;2)	(1;2;3)	(3;5;5)
8.	Dataflow complexity	0.62	[0–500]	(0;0;100)	(80;160;250)	(200;500;500)
9.	Reviews, inspection, and walkthroughs	0.61	[0–5]	(0;2;2)	(1;2;3)	(2;5;5)
10.	Software capability maturity level	0.60	[0–5]	[0–5]	(0;2;2)	(1;2;3)
11.	Minimal unit test case determination	0.59	[0–10]	(0;0;4)	(3;5;6)	(5;10;10)
12.	Requirement traceability	0.56	[0–1]	(0;0;0.4)	(0.2;0.5;0.7)	(0.5;1;1)
13.	Function point analysis	0.54	[0–100]	(0;0;5;10)	(5;10;20;30)	(25;50;100;100)
14.	System design complexity	0.53	[0–10]	(0;0;2;4)	(3;4;5;6)	(5;8;10;10)
15.	Graph-theoretic static architecture complexity	0.52	[0–10]	(0;0;2;4)	(3;4;5;6)	(5;8;10;10)
16.	Feature point analysis	0.50	[0–100]	(0;0;10)	(10;20;30)	(25;50;100;100)
17.	Requirement compliance	0.49	[0–100]	(0;0;1;5)	(3;5;10;20)	(20;40;100;100)
18.	Number of faults remaining (error seeding)	0.46	[0–100]	(0;0;5)	(5;10;15;25)	(20;40;100;100)
19.	Cause and effect graphing	0.43	[0–1]	(0;0;0.2;0.4)	(0.2;0.4;0.5;0.7)	(0.5;0.7;1;1)
20.	Cohesion	0.42	[0–1]	(0;0;0.2;0.4)	(0.2;0.4;0.5;0.7)	(0.5;0.7;1;1)
21.	Completeness	0.36	[0–1]	(0;0;0.2;0.4)	(0.2;0.4;0.5;0.7)	(0.5;0.7;1;1)

Table A.3 Metrics at coding phase with fuzzy profiles

#	Metrics	Rank	Fuzzy range	Low	Moderate	High
1.	Code defect density	0.83	[0–1]	(0;0;0.4)	(0.2;0.4;0.6)	(0.4;1;1)
2.	Design defect density	0.75	[0–5]	(0;0;2)	(1;2;3)	(2;5;5)
3.	Cyclomatic complexity	0.74	[0–500]	(0;0;150)	(100;200;300)	(250;500;500)
4.	Fault density	0.73	[0–5]	(0;0;2)	(1;2;3)	(2;5;5)
5.	Fault-days number	0.71	[0–50]	(0;0;5)	(2;7;15)	(10;50;50)
6.	Requirement specification change requests	0.69	[0–100]	(0;0;35)	(25;45;75)	(60;100;100)
7.	Error distribution	0.65	[0–10]	(0;0;3;4.5)	(3;4.5;5.5;6.5)	(5.5;8;10;10)
8.	Minimal unit test case determination	0.64	[0–10]	(0;0;2;4)	(3;5;6)	(5;8;10;10)
9.	Reviews, inspection, and walkthroughs	0.61	[0–5]	(0;2;2)	(1;2;3)	(2;5;5)
10.	Man hours per major defect detected	0.61	[0–5]	(0;0;1;2)	(1;1;2;3)	(2;3;5;5)
11.	Software capability maturity level	0.60	[0–5]	(0;2;2)	(1;2;3)	(2;5;5)
12.	Dataflow complexity	0.59	[0–500]	(0;0;100)	(80;160;250)	(200;500;500)
13.	Requirement traceability	0.56	[0–1]	(0;0;0.2;0.4)	(0.2;0.4;0.5;0.7)	(0.5;0.7;1;1)
14.	Function point analysis	0.55	[0–100]	(0;0;5;10)	(5;10;20;30)	(25;50;100;100)
15.	System design complexity	0.53	[0–10]	(0;0;2;4)	(3;4;5;6)	(5;8;10;10)
16.	Requirement compliance	0.50	[0–100]	(0;0;1;5)	(3;5;10;20)	(20;40;100;100)
17.	Feature point analysis	0.50	[0–100]	(0;0;10)	(10;20;30)	(25;50;100;100)
18.	No. of faults remaining	0.47	[0–100]	(0;0;5)	(5;10;15;25)	(20;40;100;100)
19.	Graph-theoretic static architecture complexity	0.46	[0–10]	(0;0;2;4)	(3;4;5;6)	(5;8;10;10)
20.	Bugs per line of code	0.46	[0–5]	(0;0;1;2)	(1;1;2;3)	(2;3;5;5)
21.	Cause effect graphing	0.40	[0–1]	(0;0;0.2;0.4)	(0.2;0.4;0.5;0.7)	(0.5;0.7;1;1)
22.	Cohesions	0.36	[0–1]	(0;0;0.2;0.4)	(0.2;0.4;0.5;0.7)	(0.5;0.7;1;1)
23.	Completeness	0.36	[0–1]	(0;0;0.2;0.4)	(0.2;0.4;0.5;0.7)	(0.5;0.7;1;1)

Appendix B

This appendix presents the description "qqdefects" of PROMISE repository.
Source http://promisedata.org/repository/

This is a PROMISE Software Engineering Repository dataset made publicly available in order to encourage repeatable, verifiable, refutable, and/or improvable predictive models of software engineering.

Attributes

S1   Relevant Experience of Spec and Doc Staff
S2   Quality of Documentation Inspected
S3   Regularity of Spec and Doc Reviews
S4   Standard Procedures Followed
S5   Quality of Documentation Inspected
S6   Spec Defects Discovered in Review
S7   Requirements Stability
F1   Complexity of New Functionality
F2   Scale of New Functionality Implemented
F3   Total Number of Inputs and Outputs
D1   Relevant Development Staff Experience
D2   Programmer Capability
D3   Defined Processes Followed
D4   Development Staff Motivation
T1   Testing Process Well Defined
T2   Testing Staff Experience
T3   Testing Staff Experience
T4   Quality of Documented Test Cases
P1   Development Staff Training Quality
P2   Requirements Management
P3   Project Planning
P4   Scale of Distributed Communication
P5   Stakeholder Involvement

A. K. Pandey and N. K. Goyal, *Early Software Reliability Prediction*, 135
Studies in Fuzziness and Soft Computing 303, DOI: 10.1007/978-81-322-1176-1,
© Springer India 2013

P6   Stakeholder Involvement
P7   Vendor Management
P8   Internal Communication/Interaction
P9   Process Maturity
E   Total Effort
K   KLOC
L   Language
TD   Testing Defects

Promise Dataset

1. H,M,VH,H,M,H,L,M,L,M,L,H,H,H,M,H,L,H,VH,H,H,L,H,M,
   ?,VH,H,7108.82,6.02,C,148
2. H,H,VH,H,M,H,H,L,VL,M,L,H,H,H,H,H,L,H,VH,H,H,L,H,
   M,?,VH,H,1308.08,0.897,C,31
3. H,H,VH,H,H,VH,H,H,H,VH,H,VH,H,VH,H,H,H,H,H,VH,H,?,
   VH,VH,?,VH,VH,18170,53.858,C,209
4. L,L,M,L,L,L,L,M,L,M,L,M,L,M,VL,VL,VL,L,L,M,VL,L,M,
   M,M,H,M,7006,?,C,228
5. H,M,H,M,H,?,M,H,H,VH,L,M,H,H,M,M,L,M,H,H,H,M,M,H,L,
   VH,M,9434,14,C,373
6. VH,M,VH,M,H,?,H,M,M,VH,M,H,M,M,H,?,M,M,H,H,H,M,M,
   VH,L,VH,H,9440.95,14,C,167
7. L,M,VH,H,H,L,M,L,VL,M,M,VH,H,H,H,M,M,H,H,H,VH,VL,
   VH,VH,?,H,VH,13888.26667,21,C,204
8. M,M,H,M,H,L,H,M,L,M,H,H,M,M,H,M,M,M,M,H,H,VL,H,H,?,
   H,H,8822,5.794,C,53
9. H,VH,VH,H,VH,M,VH,L,L,M,H,VH,VH,H,H,VH,VH,H,VH,VH,
   VH,L,VH,VH,?,VH,VH,2192,2.474,VC++MFC,17
10. H,H,H,M,H,M,H,M,L,M,H,H,H,H,H,M,M,M,M,H,H,H,VL,H,H,
    ?,M,H,4410,4.843,C,29
11. H,M,H,M,H,H,H,H,H,H,H,H,H,H,H,H,H,M,M,H,H,H,VL,H,H,
    ?,M,M,14196,4.371,C,71
12. H,M,H,M,M,M,L,H,H,H,VH,M,M,H,H,H,M,M,H,H,H,L,H,H,
    ?,M,H,13387.5,18.995,C,90
13. VH,M,M,L,M,H,L,H,H,H,H,H,H,M,M,M,L,M,M,H,H,VL,H,
    M,H,M,M,25449.6,49.097,C,129
14. H,H,H,H,H,H,H,VH,H,H,H,H,H,H,H,H,H,H,H,H,H,?,H,H,
    ?,H,H,33472,58.3,C,672
15. H,H,H,H,H,VH,VL,H,H,M,H,H,H,H,M,H,M,M,VH,M,H,M,VH,
    VH,?,VH,H,34892.65,154,C,1768
16. H,H,H,H,H,H,M,L,VL,M,H,H,H,H,M,H,M,M,VH,M,H,M,VH,
    VH,?,VH,H,7121,26.67,C,109

17. VH,H,M,L,H,H,M,L,VL,M,M,M,H,H,M,L,L,H,M,M,M,M,M,
    H,?,H,M,13680,33,C,688

18. M,H,H,H,H,VH,VL,VH,VH,H,M,H,H,H,H,H,M,M,VH,M,H,H,
    VH,VH,,VH,H,32365.98,155.2,C,1906

19. H,M,H,H,H,H,M,H,H,H,H,H,H,H,H,H,M,M,H,M,H,H,L,H,H,?,
    H,H,12387.65,87,C,476

20. L,L,M,VL,L,M,VL,VH,H,VH,VL,VL,L,H,VL,VL,VH,H,H,M,
    L,H,H,M,?,H,H,52660,50,C,928

21. H,H,H,M,L,M,M,L,M,VH,H,H,H,H,H,H,H,H,H,H,H,H,H,
    H,?,H,H,18748,22,C,196

22. L,L,M,M,M,M,L,M,M,VH,H,M,L,H,H,M,M,H,H,H,M,H,H,
    H,?,H,H,28206,44,C,184

23. M,H,VH,H,L,M,M,H,VH,VH,L,H,H,H,H,H,H,H,H,H,H,H,H,
    M,?,H,H,53995,61,C,680

24. M,M,M,H,M,H,L,M,M,H,M,H,H,M,H,M,M,M,H,H,M,L,M,
    H,?,VH,H,24895,99,C,1597

25. M,H,?,H,M,M,M,H,VL,H,M,H,M,H,VL,M,H,L,M,M,M,M,M,M,
    M,H,H,6905.75,23,C,546

26. M,M,H,M,H,H,H,M,H,M,L,M,M,M,M,L,H,M,L,M,M,L,H,H,
    L,H,M,1642,?,C,261

27. H,M,VH,M,M,VH,M,H,VH,VH,M,L,M,H,M,M,M,M,M,H,M,L,L,
    M,H,H,H,M,14602,52,C,412

28. H,L,VH,M,M,M,L,VH,VH,VH,M,L,H,H,M,M,M,M,H,M,L,L,
    M,M,?,H,M,8581,36,C,881

29. H,M,VH,H,M,M,VH,M,M,H,VH,VH,H,H,H,VH,VH,H,M,H,H,
    L,VH,VH,?,H,H,3764,11,C,91

30. H,H,VH,H,H,M,VH,L,L,M,H,H,H,H,H,H,H,H,M,H,H,L,VH,
    VH,?,H,H,1976,1,C,5

31. ?,H,H,M,M,H,M,M,M,H,H,H,H,H,M,H,H,M,H,H,H,H,
    VH,VH,?,VH,VH,15691,33,C,653

VL Very low, *L* Low, *M* Medium, *H* High, *VH* Very high

Appendix C

The appendix presents the description of KC2 data of NASA (2004) and can be availed from http://mdp.ivv.nasa.gov/

#	LOC	CC	EC	DC	EL	CL	BL	CCL	n1	n2	N1	N2	BC	FP
1	1.1	1.4	1.4	1.4	2	2	2	2	1.2	1.2	1.2	1.2	1.4	No
2	1	1	1	1	1	1	1	1	1	1	1	1	1	Yes
3	415	59	50	51	359	35	9	10	47	106	692	467	106	Yes
4	230	33	10	16	174	15	34	5	23	67	343	232	65	Yes
5	175	26	12	13	142	7	19	4	18	58	310	190	51	Yes
6	163	16	13	11	139	2	20	0	19	53	260	180	31	Yes
7	152	11	6	11	114	18	17	0	18	50	256	176	21	Yes
8	3	1	1	1	1	0	0	0	1	0	1	0	1	No
9	14	2	1	2	8	0	1	0	9	7	13	9	3	No
10	10	2	1	2	8	0	0	0	8	4	11	7	3	No
11	8	1	1	1	3	0	0	1	4	5	5	5	1	No
12	6	1	1	1	2	0	2	0	3	2	3	2	1	No
13	14	2	1	1	3	9	0	0	6	4	6	5	3	No
14	4	1	1	1	2	0	0	0	3	4	5	4	1	No
15	2	1	1	1	0	0	0	0	3	2	3	2	1	No
16	2	1	1	1	0	0	0	0	1	0	1	0	1	No
17	39	4	1	2	29	1	7	0	19	17	49	34	7	No
18	5	1	1	1	2	1	0	0	5	2	5	2	1	No
19	9	2	1	1	8	0	0	0	8	10	11	10	3	No
20	9	2	1	1	8	0	0	0	8	10	11	10	3	No
21	6	1	1	1	5	0	0	0	5	7	8	8	1	No
22	6	1	1	1	5	0	0	0	5	7	8	8	1	No
23	11	1	1	1	8	0	1	0	7	13	11	14	1	No
24	14	2	1	2	11	1	0	0	11	11	17	16	3	No
25	63	10	4	5	55	2	4	0	14	25	92	65	19	No
26	3	1	1	1	1	0	0	0	1	0	1	0	1	No
27	74	12	4	5	56	5	8	0	22	37	126	67	19	No
28	10	1	1	1	5	0	0	0	3	8	9	8	1	No
29	7	1	1	1	5	0	0	0	3	6	9	8	1	No
30	4	1	1	1	2	0	0	0	5	1	5	1	1	No

(continued)

A. K. Pandey and N. K. Goyal, *Early Software Reliability Prediction*,
Studies in Fuzziness and Soft Computing 303, DOI: 10.1007/978-81-322-1176-1,
© Springer India 2013

Appendix C (continued)

#	LOC	CC	EC	DC	EL	CL	BL	CCL	n1	n2	N1	N2	BC	FP
31	5	1	1	1	2	0	0	0	3	2	3	2	1	No
32	38	9	5	5	34	1	1	0	13	19	58	40	17	No
33	15	3	1	2	12	0	1	0	10	10	19	13	5	No
34	45	6	1	3	35	1	5	1	16	23	62	42	11	No
35	39	4	1	2	29	1	4	2	12	19	61	44	7	No
36	7	2	1	2	5	0	0	0	6	2	7	3	3	No
37	82	10	1	10	73	0	7	0	11	30	138	82	19	No
38	13	2	1	2	8	0	1	0	8	5	11	7	3	No
39	12	3	1	2	9	0	1	0	9	9	17	12	5	No
40	64	4	1	3	28	28	5	0	9	24	54	38	7	No
41	4	1	1	1	2	0	0	0	3	1	3	1	1	No
42	4	1	1	1	2	0	0	0	3	1	3	1	1	No
43	4	1	1	1	2	0	0	0	3	1	3	1	1	No
44	4	1	1	1	2	0	0	0	3	1	3	1	1	No
45	4	1	1	1	2	0	0	0	3	1	3	1	1	No
46	4	1	1	1	2	0	0	0	3	1	3	1	1	No
47	4	1	1	1	2	0	0	0	3	1	3	1	1	No
48	9	1	1	1	4	0	2	0	4	3	5	3	1	No
49	64	6	1	6	49	7	5	0	14	25	116	69	11	No
50	4	2	1	1	2	0	0	0	5	5	6	5	3	No
51	12	1	1	1	6	0	2	0	10	4	18	5	1	No
52	23	3	1	2	11	6	4	0	11	13	24	16	5	No
53	20	4	1	3	14	3	1	0	10	12	29	16	7	No
54	13	1	1	1	4	0	2	2	3	7	7	7	1	No
55	15	1	1	1	9	0	3	0	11	8	21	11	1	No
56	13	1	1	1	6	3	2	0	10	7	21	11	1	No
57	26	4	1	3	19	0	3	0	13	13	35	18	7	No
58	6	1	1	1	4	0	0	0	7	2	9	2	1	No
59	15	2	1	1	10	0	1	0	12	10	25	13	3	No
60	6	1	1	1	4	0	0	0	7	2	9	2	1	No
61	4	1	1	1	2	0	0	0	6	1	8	1	1	No
62	6	2	1	1	4	0	0	0	6	7	8	7	3	No
63	3	1	1	1	1	0	0	0	1	0	1	0	1	No
64	17	3	1	2	13	0	2	0	14	15	41	23	5	No
65	22	2	1	2	8	1	4	0	10	14	34	24	1	No
66	16	1	1	1	13	2	5	0	9	9	22	12	3	No
67	23	2	1	2	18	0	2	0	11	18	50	34	3	No
68	28	2	1	2	21	1	3	0	11	26	82	48	3	No
69	9	1	1	1	2	0	0	0	3	2	18	11	1	No
70	7	1	1	1	5	0	0	0	2	10	8	11	1	No
71	4	1	1	1	4	0	0	0	1	2	1	3	1	No
72	4	1	1	1	2	0	0	0	3	1	3	1	1	No
73	4	1	1	1	2	0	0	0	4	2	4	2	1	No
74	5	1	1	1	2	0	1	0	3	1	3	1	1	No

(continued)

Appendix C (continued)

#	LOC	CC	EC	DC	EL	CL	BL	CCL	n1	n2	N1	N2	BC	FP
75	15	1	1	1	12	0	1	0	6	11	34	21	1	No
76	4	1	1	1	2	0	0	0	3	2	3	2	1	No
77	56	3	1	2	78	3	13	0	11	34	105	57	17	No
78	4	1	1	1	2	0	0	0	4	3	4	3	1	No
79	7	1	1	1	2	2	1	0	5	0	5	0	1	No
80	78	6	5	6	67	5	4	0	16	34	191	111	11	No
81	3	1	1	1	1	0	0	0	1	0	1	0	1	No
82	16	2	1	2	12	0	2	0	9	5	15	8	3	No
83	39	4	3	3	30	2	5	0	15	16	47	28	7	No
84	30	6	5	4	20	2	5	0	14	14	44	28	11	No
85	4	1	1	1	2	0	0	0	3	1	3	1	1	No
86	102	25	4	4	92	3	5	0	17	47	142	106	49	No
87	10	2	1	2	8	0	0	0	8	6	13	9	3	No
88	40	4	1	4	29	1	7	1	11	18	64	38	7	No
89	46	4	1	4	35	1	7	1	13	18	71	43	7	No
90	29	6	4	4	25	0	2	0	12	13	40	27	11	No
91	3	1	1	1	1	0	0	0	1	0	1	0	1	No
92	68	5	1	3	25	0	4	0	12	24	94	73	9	No
93	31	5	1	3	51	0	14	0	11	13	47	29	9	No
94	4	1	1	1	2	0	0	0	3	1	3	1	1	No
95	16	1	1	1	5	0	0	0	4	5	4	7	1	No
96	8	1	1	1	1	0	0	0	4	4	4	6	1	No
97	7	1	1	1	5	0	0	0	2	12	10	13	1	No
98	4	1	1	1	4	0	0	0	1	0	1	0	1	No
99	45	5	1	4	37	1	5	0	16	25	74	43	9	No
100	4	1	1	1	2	0	0	0	5	1	5	1	1	No
101	4	1	1	1	2	0	0	0	3	1	3	1	1	No
102	17	6	1	5	13	0	2	0	8	12	28	22	11	No
103	4	1	1	1	2	0	0	0	3	1	3	1	1	No
104	4	1	1	1	2	0	0	0	3	1	3	1	1	No
105	4	1	1	1	2	0	0	0	3	1	3	1	1	No
106	6	1	1	1	3	0	1	0	3	2	3	2	1	No
107	15	6	1	5	13	0	0	0	6	11	28	16	11	No
108	4	1	1	1	2	0	0	0	3	1	3	1	1	No
109	4	1	1	1	2	0	0	0	3	1	3	1	1	No
110	4	1	1	1	2	0	0	0	3	2	3	2	1	No
111	4	1	1	1	2	0	0	0	5	1	5	1	1	No
112	188	8	4	5	136	9	40	0	19	91	326	183	15	No
113	4	1	1	1	2	0	0	0	3	2	3	2	1	No
114	8	1	1	1	4	0	1	0	5	3	6	4	1	No
115	4	1	1	1	2	0	0	0	3	2	3	2	1	No
116	4	1	1	1	2	0	0	0	3	2	3	2	1	No
117	8	1	1	1	4	0	1	0	5	3	6	4	1	No
118	8	1	1	1	4	0	1	0	5	3	6	4	1	No

(continued)

Appendix C (continued)

#	LOC	CC	EC	DC	EL	CL	BL	CCL	n1	n2	N1	N2	BC	FP
119	20	2	1	2	16	0	1	0	9	9	21	12	3	No
120	7	1	1	1	4	0	0	0	5	3	6	4	1	No
121	8	1	1	1	4	0	1	0	5	3	6	4	1	No
122	8	1	1	1	4	0	1	0	5	3	6	4	1	No
123	9	1	1	1	4	0	2	0	5	3	6	4	1	No
124	8	1	1	1	4	0	1	0	5	3	6	4	1	No
125	5	1	1	1	2	0	1	0	3	2	3	2	1	No
126	4	1	1	1	2	0	0	0	3	2	3	2	1	No
127	8	1	1	1	4	0	1	0	5	3	6	4	1	No
128	8	1	1	1	4	0	1	0	5	3	6	4	1	No
129	4	1	1	1	2	0	0	0	3	2	3	2	1	No
130	82	10	4	5	59	2	19	0	17	29	108	73	19	No
131	38	2	1	2	25	0	2	0	9	26	40	29	3	No
132	57	4	1	3	43	0	12	0	17	27	80	56	7	No
133	31	5	1	3	25	0	4	0	11	19	43	36	9	No
134	7	1	1	1	4	0	1	0	5	2	6	3	1	No
135	2	1	1	1	0	0	0	0	1	0	1	0	1	No
136	5	1	1	1	3	0	0	0	1	2	1	3	1	No
137	4	1	1	1	2	0	0	0	3	1	3	1	1	No
138	4	1	1	1	2	0	0	0	3	1	3	1	1	No
139	4	1	1	1	2	0	0	0	3	1	3	1	1	No
140	28	4	1	4	22	1	3	0	10	7	40	19	7	No
141	4	1	1	1	2	0	0	0	3	2	3	2	1	No
142	4	1	1	1	2	0	0	0	5	1	5	1	1	No
143	26	4	1	4	19	0	5	0	9	6	36	15	7	No
144	3	1	1	1	1	0	0	0	1	0	1	0	1	No
145	11	2	1	2	8	0	1	0	8	4	11	6	3	No
146	4	1	1	1	2	0	0	0	3	2	3	2	1	No
147	5	1	1	1	2	0	0	0	5	0	5	0	1	No
148	56	4	1	4	41	3	3	0	15	20	64	38	7	No
149	56	4	1	4	40	3	3	0	15	19	64	37	7	No
150	21	2	1	2	16	0	3	0	12	7	26	10	3	No
151	21	2	1	2	15	1	2	0	10	7	23	10	3	No
152	34	3	1	3	19	7	5	0	15	11	33	20	5	No
153	47	5	4	4	19	3	5	0	16	23	69	49	5	No
154	30	3	3	2	35	3	7	0	14	11	38	21	9	No
155	43	4	1	4	28	7	6	0	18	17	55	34	7	No
156	38	3	1	3	32	1	2	0	15	27	76	58	5	No
157	4	1	1	1	3	0	0	0	2	4	2	8	1	No
158	4	1	1	1	2	0	0	0	1	3	1	7	1	No
159	3	1	1	1	2	0	0	0	1	2	1	6	1	No
160	45	3	1	3	28	10	5	0	16	15	52	32	5	No
161	45	4	1	4	30	7	4	0	13	15	57	31	7	No
162	4	1	1	1	2	0	0	0	3	1	3	1	1	No

(continued)

Appendix C (continued)

#	LOC	CC	EC	DC	EL	CL	BL	CCL	n1	n2	N1	N2	BC	FP
163	4	1	1	1	2	0	0	0	3	1	3	1	1	No
164	4	1	1	1	2	0	0	0	6	0	6	0	1	No
165	4	1	1	1	2	0	0	0	3	1	3	1	1	No
166	4	1	1	1	2	0	0	0	3	1	3	1	1	No
167	4	1	1	1	2	0	0	0	3	1	3	1	1	No
168	5	1	1	1	2	0	0	0	6	2	6	2	1	No
169	11	2	1	2	9	0	0	0	8	6	17	10	3	No
170	47	2	1	2	119	6	18	0	11	25	87	44	31	No
171	28	2	1	2	18	1	7	0	11	15	41	28	3	No
172	8	1	1	1	4	0	1	0	3	6	7	6	1	No
173	3	1	1	1	1	0	0	0	1	0	1	0	1	No
174	97	8	7	8	72	1	20	0	14	35	147	98	15	No
175	13	2	1	2	8	0	1	0	9	6	12	8	3	No
176	10	2	1	2	8	0	0	0	8	4	11	7	3	No
177	7	1	1	1	3	0	0	1	5	5	6	5	1	No
178	31	3	1	3	23	1	4	0	11	14	47	28	5	No
179	15	3	1	2	12	0	1	0	10	10	32	17	5	No
180	21	5	1	3	18	0	1	0	11	13	43	25	9	No
181	8	1	1	1	2	0	0	0	9	9	20	13	1	No
182	4	1	1	1	4	1	0	1	4	6	7	6	1	No
183	15	3	1	2	10	0	3	0	10	10	32	17	5	No
184	14	3	1	2	10	0	1	1	10	10	32	17	5	No
185	21	4	1	4	15	0	0	0	10	15	41	28	7	No
186	17	4	1	2	18	1	0	0	7	12	25	20	7	No
187	6	1	1	1	4	0	0	0	4	6	8	7	1	No
188	6	1	1	1	4	0	0	0	4	6	8	7	1	No
189	6	1	1	1	4	0	0	0	4	6	8	7	1	No
190	6	1	1	1	4	0	0	0	4	6	8	7	1	No
191	6	1	1	1	4	0	0	0	4	6	8	7	1	No
192	6	1	1	1	4	0	0	0	4	6	8	7	1	No
193	6	1	1	1	4	0	0	0	4	6	8	7	1	No
194	6	1	1	1	4	0	0	0	4	6	8	7	1	No
195	41	8	7	7	31	1	5	2	18	16	67	38	15	No
196	56	7	3	6	49	1	4	0	22	27	86	55	13	No
197	60	7	4	6	54	1	3	0	22	30	108	65	13	No
198	55	12	9	9	45	3	5	0	11	30	113	74	23	No
199	4	1	1	1	2	0	0	0	3	1	3	1	1	No
200	33	6	4	4	26	2	3	0	16	14	52	32	11	No
201	58	13	3	5	36	10	8	2	14	28	93	70	25	No
202	19	3	3	2	14	1	2	0	11	7	21	14	5	No
203	8	2	1	2	6	0	0	0	6	7	13	8	3	No
204	18	2	1	2	13	1	2	0	10	12	24	18	3	No
205	26	5	1	2	24	0	0	0	11	10	31	21	9	No
206	29	2	1	2	24	0	2	0	10	29	60	45	3	No

(continued)

Appendix C (continued)

#	LOC	CC	EC	DC	EL	CL	BL	CCL	n1	n2	N1	N2	BC	FP
207	16	4	3	2	13	0	0	1	13	16	28	21	7	No
208	6	1	1	1	3	0	0	0	5	6	8	6	1	No
209	29	3	1	3	23	2	2	0	9	10	48	24	5	No
210	4	1	1	1	2	0	0	0	3	1	3	1	1	No
211	4	1	1	1	2	0	0	0	3	1	3	1	1	No
212	2	1	1	1	0	0	0	0	3	1	3	1	1	No
213	4	1	1	1	2	0	0	0	3	1	3	1	1	No
214	4	1	1	1	2	0	0	0	7	0	9	0	1	No
215	2	1	1	1	0	0	0	0	3	1	3	1	1	No
216	4	1	1	1	2	0	0	0	3	1	3	1	1	No
217	4	1	1	1	2	0	0	0	3	1	3	1	1	No
218	4	1	1	1	2	0	0	0	3	1	3	1	1	No
219	4	1	1	1	2	0	0	0	3	1	3	1	1	No
220	4	1	1	1	2	0	0	0	3	1	3	1	1	No
221	4	1	1	1	2	0	0	0	3	1	3	1	1	No
222	4	1	1	1	2	0	0	0	3	1	3	1	1	No
223	4	1	1	1	2	0	0	0	3	1	3	1	1	No
224	6	3	1	1	4	0	0	0	6	7	13	11	5	No
225	4	1	1	1	2	0	0	0	3	1	3	1	1	No
226	4	1	1	1	2	0	0	0	3	1	3	1	1	No
227	4	1	1	1	2	0	0	0	3	1	3	1	1	No
228	4	1	1	1	2	0	0	0	3	1	3	1	1	No
229	4	1	1	1	2	0	0	0	3	1	3	1	1	No
230	4	1	1	1	2	0	0	0	3	1	3	1	1	No
231	4	1	1	1	2	0	0	0	4	1	4	1	1	No
232	4	1	1	1	2	0	0	0	3	1	3	1	1	No
233	4	1	1	1	2	0	0	0	3	1	3	1	1	No
234	4	1	1	1	2	0	0	0	3	1	3	1	1	No
235	6	3	1	1	4	0	0	0	6	7	13	11	5	No
236	4	1	1	1	2	0	0	0	7	0	9	0	1	No
237	4	1	1	1	2	0	0	0	3	1	3	1	1	No
238	4	1	1	1	2	0	0	0	3	1	3	1	1	No
239	6	1	1	1	3	0	1	0	3	2	3	2	1	No
240	4	1	1	1	2	0	0	0	3	1	3	1	1	No
241	4	1	1	1	2	0	0	0	3	1	3	1	1	No
242	20	2	1	2	14	3	1	0	6	10	33	21	3	No
243	6	1	1	1	3	0	1	0	3	2	3	2	1	No
244	36	8	8	5	26	1	6	1	18	16	72	41	15	No
245	4	1	1	1	2	0	0	0	5	2	5	2	1	No
246	2	1	1	1	0	0	0	0	7	1	7	1	1	No
247	4	1	1	1	2	0	0	0	3	1	3	1	1	No
248	4	1	1	1	2	0	0	0	4	1	4	1	1	No
249	29	3	1	3	22	0	4	1	11	11	54	29	5	No
250	4	1	1	1	2	0	0	0	6	2	6	2	1	No

(continued)

Appendix C (continued)

#	LOC	CC	EC	DC	EL	CL	BL	CCL	n1	n2	N1	N2	BC	FP
251	44	9	4	4	29	4	8	1	12	25	77	60	17	No
252	4	1	1	1	2	0	0	0	3	1	3	1	1	No
253	4	1	1	1	2	0	0	0	3	2	3	2	1	No
254	4	1	1	1	2	0	0	0	3	2	3	2	1	No
255	2	1	1	1	0	0	0	0	3	2	3	2	1	No
256	4	1	1	1	2	0	0	0	3	2	3	2	1	No
257	4	1	1	1	2	0	0	0	3	2	3	2	1	No
258	4	1	1	1	2	0	0	0	3	2	3	2	1	No
259	5	1	1	1	2	0	0	0	3	2	3	2	1	No
260	7	1	1	1	4	0	1	0	4	3	5	3	1	No
261	7	4	1	1	5	0	0	0	6	9	17	16	7	No
262	138	20	18	15	128	0	8	0	16	44	275	169	39	No
263	4	1	1	1	2	0	0	0	3	2	3	2	1	No
264	2	1	1	1	0	0	0	0	3	2	3	2	1	No
265	10	1	1	1	5	0	3	0	6	3	8	4	1	No
266	4	1	1	1	2	0	0	0	3	2	3	2	1	No
267	21	2	1	2	16	0	3	0	10	13	40	22	3	No
268	4	1	1	1	2	0	0	0	3	2	3	2	1	No
269	4	1	1	1	2	0	0	0	3	2	3	2	1	No
270	7	4	1	1	5	0	0	0	6	9	17	16	7	No
271	20	2	1	2	16	0	1	0	9	8	21	11	3	No
272	4	1	1	1	2	0	0	0	3	2	3	2	1	No
273	5	1	1	1	2	1	0	0	4	1	4	1	1	No
274	33	4	3	4	24	3	3	1	11	18	55	29	7	No
275	4	1	1	1	2	0	0	0	4	1	4	1	1	No
276	94	12	5	7	69	9	12	2	19	44	142	108	23	No
277	5	1	1	1	2	1	0	0	4	1	4	1	1	No
278	8	1	1	1	3	0	2	0	4	3	5	3	1	No
279	12	2	1	2	6	3	1	0	6	6	9	6	3	No
280	12	2	1	2	8	1	0	0	6	2	11	4	3	No
281	5	1	1	1	2	0	0	0	6	2	6	2	1	No
282	5	1	1	1	2	0	0	0	5	0	5	0	1	No
283	51	6	1	5	35	2	11	0	16	18	79	46	11	No
284	10	2	1	1	6	0	2	0	8	9	16	12	3	No
285	88	8	4	6	67	3	11	0	13	43	143	101	15	No
286	47	7	6	6	48	3	6	1	21	20	77	36	15	No
287	44	6	6	6	35	1	5	0	21	21	80	44	11	No
288	41	7	1	6	33	2	3	1	18	16	61	32	13	No
289	21	3	3	2	16	0	3	0	15	7	28	14	5	No
290	6	2	1	1	10	0	2	0	6	9	10	9	3	No
291	4	1	1	1	2	0	0	0	7	0	9	0	1	No
292	2	1	1	1	0	0	0	0	3	1	3	1	1	No

(continued)

Appendix C (continued)

#	LOC	CC	EC	DC	EL	CL	BL	CCL	n1	n2	N1	N2	BC	FP
293	4	1	1	1	2	0	0	0	4	2	4	2	1	No
294	4	1	1	1	2	0	0	0	3	1	3	1	1	No
295	4	1	1	1	2	0	0	0	3	1	3	1	1	No
296	4	1	1	1	2	0	0	0	3	1	3	1	1	No
297	10	5	1	1	6	0	1	0	7	5	23	10	9	No
298	20	2	1	2	15	0	2	0	10	13	33	21	3	No
299	241	20	4	20	161	43	34	0	28	61	381	238	39	No
300	2	1	1	1	0	0	0	0	3	2	3	2	1	No
301	66	13	3	7	52	4	7	1	19	23	103	58	25	No
302	4	1	1	1	2	0	0	0	4	3	4	3	1	No
303	4	1	1	1	2	0	0	0	3	2	3	2	1	No
304	10	2	1	1	6	0	2	0	8	8	14	10	3	No
305	11	1	1	1	4	3	2	0	5	2	11	2	1	No
306	12	2	1	2	8	0	1	0	8	4	12	6	3	No
307	144	14	7	11	106	29	7	0	25	43	205	117	27	No
308	11	2	1	2	8	0	1	0	8	4	12	7	3	No
309	22	3	1	3	16	2	2	0	9	12	31	19	5	No
310	11	2	1	2	8	0	1	0	8	4	11	6	3	No
311	10	2	1	1	8	0	0	0	7	4	10	7	3	No
312	30	2	1	2	19	1	8	0	11	14	38	25	3	No
313	43	4	3	4	30	1	9	1	10	16	56	37	7	No
314	83	7	1	6	66	1	12	0	14	23	135	88	13	No
315	4	1	1	1	2	0	0	0	4	1	4	1	1	No
316	106	12	7	8	81	5	15	1	14	34	157	104	23	No
317	3	1	1	1	1	0	0	0	1	0	1	0	1	No
318	11	2	1	2	8	0	1	0	8	5	12	7	3	No
319	11	2	1	2	8	0	1	0	8	4	12	7	3	No
320	4	1	1	1	2	0	0	0	1	1	1	1	1	No
321	6	1	1	1	2	0	2	0	3	1	3	1	1	No
322	18	2	1	2	12	0	2	0	9	8	27	14	3	No
323	20	3	3	3	5	0	0	0	11	8	23	15	5	No
324	7	3	1	1	14	0	1	0	8	5	11	7	5	No
325	100	9	1	8	73	5	17	0	19	41	144	95	17	No
326	28	2	1	2	21	0	5	0	10	15	35	25	3	No
327	10	2	1	2	8	0	0	0	8	4	11	7	3	No
328	30	5	3	4	20	3	5	0	12	18	48	38	9	No
329	20	4	3	3	16	0	2	0	15	7	26	14	7	No
330	23	3	1	3	18	0	3	0	13	13	31	20	5	No
331	12	2	1	2	7	0	3	0	9	5	11	9	3	No
332	20	3	1	2	15	0	3	0	11	12	29	17	5	No
333	28	2	1	2	18	0	6	0	10	12	35	24	3	No
334	28	2	1	2	18	0	6	0	10	12	35	24	3	No

(continued)

Appendix C (continued)

#	LOC	CC	EC	DC	EL	CL	BL	CCL	n1	n2	N1	N2	BC	FP
335	3	1	1	1	1	0	0	0	1	0	1	0	1	No
336	78	5	1	5	52	6	17	1	12	20	106	66	9	No
337	14	2	1	2	8	0	1	0	9	7	13	9	3	No
338	10	2	1	2	8	0	0	0	8	4	11	7	3	No
339	8	1	1	1	3	0	0	1	4	5	5	5	1	No
340	29	3	1	3	22	1	3	0	11	14	45	27	5	No
341	3	1	1	1	1	0	0	0	1	0	1	0	1	No
342	21	2	1	2	16	1	1	0	10	8	23	11	3	No
343	13	2	1	2	8	1	2	0	8	4	12	6	3	No
344	43	5	1	4	30	3	7	0	15	19	61	40	9	No
345	12	2	1	2	7	1	2	0	7	4	10	5	3	No
346	18	5	1	2	12	1	3	0	11	9	32	22	9	No
347	4	1	1	1	1	0	0	0	1	0	1	0	1	No
348	3	1	1	1	1	0	0	0	1	0	1	0	1	No
349	7	1	1	1	3	0	1	0	5	1	5	1	1	No
350	33	6	1	2	18	3	9	0	11	15	48	35	11	No
351	43	3	1	3	29	4	6	0	16	12	47	25	5	No
352	7	1	1	1	3	0	1	0	5	1	5	1	1	No
353	52	4	1	3	37	5	6	0	18	19	63	36	7	No
354	48	3	1	3	34	5	5	0	16	13	55	29	5	No
355	47	3	1	3	33	5	5	0	16	13	52	28	5	No
356	30	5	1	4	23	0	5	0	12	16	37	21	9	No
357	91	8	6	7	74	0	15	0	19	31	133	88	15	No
358	70	6	1	4	52	3	9	1	16	28	105	70	11	No
359	28	5	1	2	18	2	6	0	13	18	46	32	9	No
360	4	1	1	1	2	0	0	0	3	1	3	1	1	No
361	108	11	1	8	85	0	20	0	15	38	172	117	21	No
362	4	2	1	1	2	0	0	0	5	6	7	6	3	No
363	5	1	1	1	2	0	0	1	3	2	3	2	1	No
364	174	29	7	14	133	4	32	2	23	56	272	194	57	No
365	100	8	1	8	82	2	13	0	13	33	176	111	15	No
366	64	5	1	5	50	0	7	0	12	28	116	73	9	No
367	55	13	1	13	40	0	13	0	13	20	67	33	25	No
368	59	5	1	4	51	1	5	0	17	30	78	68	9	No
369	71	11	1	6	58	0	11	0	16	22	104	67	21	No
370	30	6	1	3	20	3	4	1	8	13	37	26	11	No
371	33	6	1	3	24	0	7	0	9	16	44	31	11	No
372	49	4	1	2	19	18	7	0	15	20	38	39	7	No
373	4	1	1	1	2	0	0	0	3	1	3	1	1	No
374	29	5	1	3	22	0	5	0	11	17	40	30	9	No
375	50	3	1	3	39	2	4	0	13	18	80	44	5	No
376	62	6	3	5	55	1	4	0	15	32	112	72	11	No
377	19	4	1	1	15	0	1	0	9	7	26	18	7	No
378	33	2	1	2	23	0	5	0	11	15	33	20	3	No

(continued)

Appendix C (continued)

#	LOC	CC	EC	DC	EL	CL	BL	CCL	n1	n2	N1	N2	BC	FP
379	5	1	1	1	2	0	1	0	3	2	3	2	1	No
380	32	4	1	2	21	2	6	1	9	20	35	28	7	No
381	4	1	1	1	2	0	0	0	3	1	3	1	1	No
382	7	1	1	1	5	0	0	0	5	2	9	4	1	No
383	7	1	1	1	4	0	0	1	5	2	9	4	1	No
384	7	1	1	1	5	0	0	0	5	2	9	4	1	No
385	6	1	1	1	4	0	0	0	6	2	8	3	1	No
386	4	1	1	1	2	0	0	0	3	1	3	1	1	No
387	8	1	1	1	6	0	0	0	6	4	8	5	1	No
388	6	2	1	2	4	0	0	0	6	3	10	3	3	No
389	6	1	1	1	4	0	0	0	6	2	8	3	1	No
390	4	1	1	1	2	0	0	0	3	1	3	1	1	No
391	6	1	1	1	4	0	0	0	3	3	5	4	1	No
392	4	1	1	1	2	0	0	0	3	2	3	2	1	No
393	46	9	6	7	39	3	2	0	14	27	80	52	17	No
394	17	4	1	3	15	0	0	0	11	11	32	17	7	No
395	48	15	1	7	43	0	2	1	15	21	94	66	29	No
396	13	2	1	2	7	1	1	2	10	7	19	14	3	No
397	30	3	1	3	24	0	3	1	12	11	36	21	5	No
398	11	2	1	2	7	0	1	1	8	4	11	7	3	No
399	13	2	1	2	9	1	1	0	11	8	14	12	3	No
400	29	4	1	4	19	6	2	0	10	11	31	16	5	No
401	60	5	1	5	44	3	9	0	13	23	68	47	9	No
402	5	2	1	1	3	0	0	0	6	6	9	6	3	No
403	47	3	1	3	36	0	7	0	10	21	71	44	5	No
404	38	4	1	4	27	0	5	0	10	18	52	32	7	No
405	46	4	1	4	36	0	5	0	12	24	67	45	7	No
406	2	1	1	1	0	0	0	0	3	1	3	1	1	No
407	8	1	1	1	6	0	0	0	3	6	7	7	1	No
408	4	1	1	1	2	0	0	0	3	1	3	1	1	No
409	4	1	1	1	2	0	0	0	3	1	3	1	1	No
410	20	4	3	3	15	1	2	0	13	11	27	18	7	No
411	2	1	1	1	0	0	0	0	3	1	3	1	1	No
412	9	1	1	1	3	2	2	0	6	3	9	4	1	No
413	9	1	1	1	3	2	2	0	6	3	9	4	1	No
414	54	7	3	7	42	1	8	1	15	17	72	46	13	No
415	6	1	1	1	4	0	0	0	3	4	5	4	1	No
416	4	1	1	1	2	0	0	0	3	2	3	2	1	No
417	94	6	1	6	71	0	20	1	18	38	138	92	11	No
418	109	8	1	8	89	2	16	0	14	28	181	115	15	No
419	16	4	3	2	13	0	0	1	13	16	29	22	7	No
420	1	1	1	1	0	0	0	0	5	1	5	1	1	No
421	55	10	1	6	37	2	10	4	19	23	75	47	19	No
422	149	30	16	27	115	20	10	2	17	39	255	136	59	Yes

(continued)

Appendix C (continued)

#	LOC	CC	EC	DC	EL	CL	BL	CCL	n1	n2	N1	N2	BC	FP
423	133	21	9	12	98	7	20	6	20	61	239	169	41	Yes
424	130	14	6	11	104	4	18	2	14	42	204	132	27	Yes
425	120	18	16	12	107	2	9	0	14	44	211	141	35	Yes
426	1275	180	125	143	1107	39	121	6	35	325	2469	###	361	Yes
427	107	9	4	7	84	0	19	1	20	46	189	128	17	Yes
428	105	16	14	11	87	3	12	1	21	40	174	120	31	Yes
429	517	86	35	46	401	44	56	6	25	126	798	542	171	Yes
430	102	14	1	10	80	4	15	0	13	32	161	93	27	Yes
431	102	21	8	10	79	4	15	2	18	37	180	114	41	Yes
432	97	7	3	6	61	26	8	0	14	32	139	74	13	Yes
433	275	24	18	21	195	38	39	1	26	66	359	213	47	Yes
434	89	13	6	7	68	6	12	1	18	43	142	97	25	Yes
435	80	19	9	12	71	0	7	0	17	37	164	109	37	Yes
436	76	9	5	6	56	3	10	4	19	35	110	75	17	Yes
437	148	16	8	11	36	1	8	0	23	49	238	147	3	Yes
438	143	26	13	21	116	7	13	5	24	54	253	167	51	Yes
439	140	22	17	18	115	12	10	1	26	46	266	155	43	Yes
440	70	5	1	5	54	2	12	0	12	21	106	63	9	Yes
441	67	5	1	5	50	4	9	2	13	37	120	93	9	Yes
442	63	8	7	7	49	2	10	0	18	37	115	79	15	Yes
443	61	4	1	3	49	0	10	0	15	30	93	66	7	Yes
444	61	8	7	7	38	2	4	0	23	27	107	58	13	Yes
445	60	9	5	5	38	17	2	0	19	18	68	42	17	Yes
446	59	9	5	6	44	1	10	0	18	23	120	64	17	Yes
447	59	7	7	5	44	2	8	2	16	25	82	54	13	Yes
448	57	8	8	8	45	4	6	0	9	21	99	61	15	Yes
449	56	4	1	4	44	3	3	0	15	22	64	41	7	Yes
450	110	12	10	12	88	3	15	0	16	41	190	123	23	Yes
451	54	4	1	3	42	0	10	0	15	26	74	53	7	Yes
452	269	47	8	31	198	29	26	11	21	75	506	346	97	Yes
453	52	7	1	4	33	6	9	0	12	25	81	52	13	Yes
454	51	6	4	6	42	3	4	0	15	19	61	35	11	Yes
455	101	11	4	8	84	1	12	0	18	53	177	132	21	Yes
456	99	9	5	5	46	0	8	0	16	50	172	113	5	Yes
457	194	23	1	11	137	23	29	1	26	56	317	187	45	Yes
458	94	20	17	15	75	7	8	0	16	39	178	101	39	Yes
459	46	7	1	4	38	1	2	0	15	20	81	52	13	Yes
460	46	3	1	2	24	12	7	0	11	13	47	29	5	Yes
461	137	19	5	16	112	9	14	0	20	46	207	120	37	Yes
462	91	8	7	7	71	5	8	2	19	38	149	103	15	Yes
463	591	87	34	61	443	40	92	6	26	150	1147	739	173	Yes
464	45	9	6	7	39	3	1	0	14	27	80	52	17	Yes
465	42	3	1	3	31	2	5	0	13	18	64	38	5	Yes
466	42	4	1	3	23	5	9	1	14	19	44	34	7	Yes

(continued)

Appendix C (continued)

#	LOC	CC	EC	DC	EL	CL	BL	CCL	n1	n2	N1	N2	BC	FP
467	122	6	5	5	91	4	22	0	21	40	209	144	11	Yes
468	80	8	1	6	58	8	11	0	14	29	132	71	15	Yes
469	76	10	4	7	60	3	9	0	21	25	114	70	19	Yes
470	72	12	5	6	63	0	6	0	18	31	120	80	23	Yes
471	36	5	1	2	26	0	5	1	13	27	57	46	9	Yes
472	143	14	6	10	105	14	21	1	19	55	219	145	27	Yes
473	70	11	1	9	57	4	6	0	17	28	108	67	21	Yes
474	35	5	1	4	30	0	3	0	13	20	46	31	9	Yes
475	104	17	5	11	86	5	11	0	16	51	214	142	33	Yes
476	34	6	6	6	27	1	4	0	19	14	56	24	11	Yes
477	102	10	8	9	72	12	16	0	22	40	154	90	19	Yes
478	33	6	1	3	28	0	3	0	14	19	47	35	11	Yes
479	63	5	1	4	50	1	10	0	14	30	89	71	9	Yes
480	61	7	1	4	49	1	9	0	14	32	92	74	13	Yes
481	58	4	1	3	44	1	10	0	15	27	81	59	7	Yes
482	113	14	8	12	85	9	15	0	17	42	164	101	27	Yes
483	139	32	16	10	118	1	16	1	17	61	249	177	63	Yes
484	55	4	1	3	42	0	11	0	17	27	77	54	7	Yes
485	210	5	1	3	164	3	41	0	17	60	395	292	9	Yes
486	72	4	1	4	53	1	16	0	16	27	95	67	7	Yes
487	24	1	1	1	13	2	7	0	8	13	23	17	1	Yes
488	118	3	1	2	87	1	23	0	16	40	197	156	5	Yes
489	227	27	7	16	168	21	30	4	20	70	385	211	53	Yes
490	45	9	1	4	36	2	4	1	20	39	106	91	17	Yes
491	112	19	11	15	93	2	10	5	20	61	197	151	37	Yes
492	44	2	1	2	22	12	7	0	14	17	39	27	3	Yes
493	43	3	1	3	24	4	8	0	8	16	50	33	5	Yes
494	43	6	5	5	32	0	9	0	20	20	61	45	11	Yes
495	39	7	1	4	34	0	3	0	12	14	58	31	13	Yes
496	58	10	5	6	50	4	2	0	16	26	106	70	19	Yes
497	37	8	1	4	24	4	7	0	11	17	63	36	15	Yes
498	18	2	1	2	12	1	3	0	10	6	21	12	3	Yes
499	49	4	1	4	39	2	5	0	12	18	76	50	7	Yes
500	65	6	1	6	52	3	7	1	13	22	84	55	11	Yes
501	32	6	1	6	27	1	2	0	18	17	55	32	11	Yes
502	15	2	1	1	4	0	0	0	8	18	24	22	3	Yes
503	88	10	6	6	66	7	11	0	16	35	120	78	19	Yes
504	29	1	1	1	20	0	7	0	11	18	43	30	1	Yes
505	14	3	1	2	12	0	0	0	9	11	19	14	5	Yes
506	14	2	1	2	12	0	0	0	13	20	32	29	3	Yes
507	14	3	1	3	10	0	2	0	10	13	25	17	5	Yes
508	81	7	1	7	62	4	11	0	17	28	132	75	13	Yes
509	13	2	1	1	10	0	1	0	8	11	18	13	3	Yes
510	37	2	1	2	26	5	5	0	7	17	24	51	3	Yes

(continued)

Appendix C (continued)

#	LOC	CC	EC	DC	EL	CL	BL	CCL	n1	n2	N1	N2	BC	FP
511	12	2	1	1	8	0	1	0	7	10	15	12	3	Yes
512	11	3	1	2	9	0	0	0	12	10	22	12	5	Yes
513	10	2	1	2	8	0	0	0	8	4	11	7	3	Yes
514	8	1	1	1	4	0	1	0	5	3	6	4	1	Yes
515	8	1	1	1	4	0	1	0	5	3	6	4	1	Yes
516	14	3	1	1	9	0	3	0	6	10	17	15	5	Yes
517	6	2	1	1	2	0	2	0	5	7	8	7	3	Yes
518	4	1	1	1	2	0	0	0	4	1	4	1	1	Yes
519	4	1	1	1	2	0	0	0	3	1	3	1	1	Yes
520	4	1	1	1	2	0	0	0	3	1	3	1	1	Yes
521	4	1	1	1	2	0	0	0	3	2	3	2	1	Yes
522	3	1	1	1	1	0	0	0	1	0	1	0	1	Yes

About the Author

Dr. Ajeet Kumar Pandey is currently associated with AECOM India Private Limited, supporting RAM analysis of L&T Metro Rail, Hyderabad (India), one of the major public-private partnership (PPP) projects. He is also taking guest lectures, on invitation, at Indian Institute of Management Raipur, Raipur, India. He has more than 12 years of professional and research experience. He worked as a Sr. Research Fellow at IIT Kharagpur, Kharagpur, India, Sr. RAMS Engineer at Cognizant Technology Solutions, Hyderabad, India, and Sr. Systems Officer at UPTEC, Lucknow, India. Dr. Pandey received his Ph.D. in Reliability Engineering from Indian Institute of Technology Kharagpur, Kharagpur, India 2011. Prior to this, he did his M. Tech. from Motilal Nehru National Institute of Technology, Allahabad, India. His major areas of research are software engineering and project management; data mining; software metrics, software quality and testing; reliability analysis and prediction; hazard analysis and system safety; and handling compliance and regulatory issues of safety critical systems. He has also been involved with academics and has taught subjects like Operating System, Operational Research, Software Engineering, Data Mining & Business Intelligence, etc., to Undergraduate and Postgraduate students. He has also contributed several research papers to international journals and conference proceedings.

Dr. Neeraj Kumar Goyal is currently an Assistant Professor in Reliability Engineering Centre, Indian Institute of Technology Kharagpur, India. He has received his Ph.D. degree from IIT Kharagpur in Reliability Engineering in 2006. His Ph.D. topic was 'On Some Aspects of Reliability Analysis and Design of Communication Networks'. He received the Bachelor of Engineering degree with Honors in Electronics and Communications Engineering from MREC Jaipur, Rajasthan, India in 2000. His areas of research and teaching are network reliability, software reliability, electronic system reliability, reliability testing, probabilistic risk/safety assessment, and reliability design. He has completed various research and consultancy projects for various organizations, e.g. DRDO, NPCIL, Vodafone, ECIL etc. He has contributed several research papers to international journals and conference proceedings. He is also associate editor of International Journal of Performability Engineering.

A. K. Pandey and N. K. Goyal, *Early Software Reliability Prediction*,
Studies in Fuzziness and Soft Computing 303, DOI: 10.1007/978-81-322-1176-1,
© Springer India 2013

Printed in the United States
By Bookmasters